CW00495457

MOSES KELLOW

AUTOBIOGRAPHY

Moses Kellow, JP, M Inst CE, MI Mech E, MIQ, FGS, was born in 1862 and died in 1943. He was Manager of the Croesor Quarry from the mid-1890s until its closure in 1930, and beyond.

MOSES KELLOW

THE AUTOBIOGRAPHY OF THE CROESOR QUARRY MANAGER

Edited by Cathy Woodhead

EDITOR'S INTRODUCTION

As you will see from his Foreword, Adrian Barrell, for his own interest, painstakingly typed up the autobiography that Moses Kellow dictated for the Quarry Managers' Journal in the early 1940s. Adrian had a few copies of his typescript made and, by chance, I was loaned one of these whilst doing my own research into the history of the Croesor valley. The fascinating range of Kellow's life story struck me as being very suitable for publishing in book form. I contacted Adrian and the Quarry Managers' Journal (now the Quarry Management) Editor and they both gave permission to publish the material.

Adrian and other people with an interest in Croesor, Kellow and slate quarrying have been very helpful in providing additional material which has been included in the appendices. Please note that the original spellings of place names have been kept, and chapter breaks and titles have been added to Kellow's words. The maps in Appendix I may help to locate places and the route of the Croesor Railway.

In addition to Adrian Barrell I would like to thank the following people for their contributions: Cherrie Carlisle and her brother John Kellow Williams, Peredur Hughes, Richard Walwyn, Harold Morris, John Edlington, Bob Watson and Dewi Jones. I hope that the book does justice to a remarkable man and the industries that he worked in.

Cathy Woodhead
Delfryn Publications

ISBN: 978-0-9561469-9-1

Published in 2015 by Delfryn Publications, Delfryn, Borth-y-Gest, Porthmadog, Gwynedd LL49 9TW, UK. Telephone: +44(0)1766 512115

www.delfrynpublications.co.uk

Printed by Print Domain

Cover design by Cathy Woodhead from a photograph of Cnicht taken by David Medcalf

CONTENTS

FOREWORD

I first visited the quarries of Cwm Croesor in the late 1950s, arriving over Bwlch Rhosydd with a heavily laden bicycle! By the 1970s I had begun to research the history of the Park & Croesor Slate Quarries and to be intrigued by their innovative Manager and Engineer, Moses Kellow. I was told that Kellow had written his autobiography, so I began to look for a copy.

I discovered that at one time it had been the custom for the Editor of the Quarry Managers' Journal to approach senior retired quarry managers and ask for some autobiographical notes. In most cases the result was a few brief paragraphs. Moses Kellow was different! His Autobiography runs to nearly 78,000 words! This vast typescript must have caused consternation when it arrived at the offices of the Journal. The Editor decided that the best course was to serialize it, and it was published in twenty-four parts during 1944-5.

My search led me to the National Library of Wales, where I found copies of the Journal and built up my own typescript of the Autobiography. Subsequently I discovered that a carbon copy of the original is held as No.5276 in the Bangor General Collection of Manuscripts in the Library of the University College of North Wales. Until now the Autobiography has never been published in book form. Now, of course, thanks to the initiative of Delfryn Publications, Moses Kellow will be introduced to a wider public.

We have some information on the birth of the Autobiography. Kellow had retired to Cheltenham in 1935 but in 1941, wary of enemy air-raids, he returned with his wife Martha and Alice Mahon his nurse, to stay for nearly two years at Cartrefle, the Old Post Office, in the Stryd Fawr at Penrhyndeudraeth. It was here that the Autobiography was produced. Another occupant of the house at that time was a young teacher, the late Mrs Laura Rimmer. When I visited her at Cartrefle in November, 1994, she remembered the Kellow visitation very well. Her initial memory was of Martha's over-occupancy of the kitchen but, after that, she remembered the production of the Autobiography. I was shown the upstairs sitting-room in which the Autobiography had been dictated, and I am sure that it had changed very little since that time.

By 1941 Kellow was almost completely blind. Because of this he had to dictate every word to John Griffiths. "Griffiths Bach" worked full-time in the Minffordd office of Evan Morris (Morris & Co. Slate Merchants), so dictation took place during evenings and at weekends. John Griffiths would go home with the result of a dictation session and return a few days later with the typescript. It was very rare for any alterations to be made, because Moses' head was already full of what he wanted to say in the next section. He had no written sources to remind him. Everything came straight from his memory, which accounts for an occasional vagueness and for the very few dates in the Autobiography; few of us can remember any but the most significant dates. Another problem was that for some reason John Griffiths was reluctant to number the pages of the growing typescript. This led to confusion and acrimony. At last the work was finished and was sent off to the Journal. It would be very interesting to read the covering letter.

It is now over seventy years since the Autobiography was written. In the following pages you will read the words of an infirm and elderly gentleman who must have realised that his days were numbered. His still-active mind sets off on a fascinating journey through his long and varied life in times which have gone beyond human memory. Now you can share that journey with him.

Adrian Barrell
July 2015

1. MY EARLY YEARS

It is sometimes said jocularly (and I believe sometimes seriously) that a man should be very careful in choosing his parents—this topsy-turvy way of stating the matter being adopted to stress the profound influence of heredity upon the character. Well, I think I did not fail when I was selecting my parents. They were of good, healthy stock on both sides. My father had a head as large as that of Mr Gladstone, with a pronounced development on the frontal or intellectual region. He wore a hat of 7 $^5/_8$ inches.

My mother also had a large head with a good frontal development. A man once said to me, "your mother has a forehead like a field!" She was a person of intense energy, and except in one thing—where nature prescribes the pace for ladies—she did everything in her life at about double the rate of that of any other woman I have ever known, and did it efficiently. I am sufficiently democratic to prefer ancestors possessing health, brains and energy to enjoying what doubtful privileges may accrue from a lineage traceable back, as is often the case in illustrious families, to the bastard son of a profligate British King.

I first saw the light of day on the 5th of June 1862 at Old Delabole, Cornwall, famous for one of the largest Slate Quarries in the Kingdom, the pit being about 450ft deep and at least half a mile long and wide. It has been working continuously since the year 1600.

It will be seen, therefore, that I almost breathed slate from the day that I entered the world, and I am not quite sure that I was not sometimes fed upon slate and butter instead of bread and butter. At any rate, my early environment seemed to have a profound influence upon my subsequent career. It is very remarkable that almost all my relatives were slate people, and they have had considerable influence in shaping the slate industry of the world.

When I was three years of age my parents went to live at Caernarvon, north Wales, and one of my first recollections was the occasion of my paternal grandfather's funeral, to which my father took me, my grandparent having died about a year after we had moved from Cornwall. I was left behind for a short time to be a comfort to my bereaved grandmother, but the Quarry ever drew me like a magnet, and I spent most of my time there. The main haulage incline of the Delabole Quarry, extending down to a depth of about 400ft, had two roads upon it. Three wagons at a time were pulled up on one side, while three empties descended on the other. When the full loads reached the top, the procedure was reversed for the next operation, and empties were lowered on the road up which the full wagons had previously been hauled. On the other road, of course, the reverse process was taking place.

The drums were operated by a reversing steam engine which was out of sight of the incline, and inside the engine house was an indicator which rose from floor to ceiling when full wagons were coming up on the No.1 road on the incline, and descended from ceiling to floor as empties were being lowered on the same road.

I had free access to this engine room and was fascinated by the cleverness of the device which enabled the engine driver to know what was taking place on the incline. The reversing lever was pulled backwards when loads were being wound up, and pushed forward when wagons were being lowered on the same road. The indicator showed what was taking place on one of the roads, but no indicator was necessary for the second road because it was the opposite of what was taking place on the first road.

One day I was in this engine house, and observing that the operation of working this haulage engine was very simple, I was struck by an original idea. When, therefore, the full loads were half-way up the incline I threw the reversing lever over, and everyone about the incline was astonished to see the full wagons going down the incline, and the empty ones coming up. Of course, the driver promptly pulled the lever back again, and severely reprimanded me for interfering. But I assure my readers that I was a very proud little boy at having been able, at four years of age, to operate the main haulage incline of the big Delabole Quarry.

I thoroughly understood what was taking place both on the incline and in the engine house, and the description here given is founded upon a keen and clear recollection of the conditions and of what was taking place.

2. MY GUIDING PRINCIPLE

I look upon this experience as my first introduction to my engineering and quarrying career, and it first inspired in me that scientific attitude of mind which has been my guiding principle ever since.

Apropos of this, I would like to interpolate at this point an amusing recollection of about twenty years ago. I was driving my car along the road not far from the Festiniog Railway station at Penrhyndeudraeth, when I encountered a flock of sheep coming in the opposite direction. I pulled in to the left of the road and stopped the car, leaving about 10ft from the side of the car to the opposite wall, clear for the sheep to pass. Notwithstanding this ample space the whole flock stopped in line with the front of the car, but presently a ram, who was very slightly ahead of the others, suddenly stepped forward about 2ft into the 10ft gap and jumped fully 5ft high into the air over an imaginary obstacle. This set the others off, and every single sheep in that flock of about fifty came forward one by one and followed the ram's example with amazing exactitude, leaping 5ft into the air over nothing. It was one of the most ridiculous sights I have ever seen, as the jumping took place on my right-hand side and I got a perfect view of the performance.

Now this shows the tremendous power of the instinct to imitate, and this instinct is not only to be found in animals but also in man, and few people realise to what an astonishing extent it is responsible for people's actions in the mass, and for the views they hold.

If someone, often of little importance or significance, does a thing, a whole crowd will emulate him. Likewise, let someone prominent in the public eye make a statement, and however unfounded on knowledge or fact it may be, the majority will accept it as right without ever dreaming of criticising it, much less probing its validity.

For many years workmen from the Festiniog Quarries, living at Penrhyndeudraeth, would, on emerging from the train at their home station, immediately begin to run or trot downhill towards the town, and about a quarter of a mile of roadway would be filled with running men, thinning out as groups would turn from the main group, in order, still trotting, to take the nearest side streets to their respective homes.

I suppose that about 40 years previously one man was in a hurry and ran all the way home. I can well imagine the result. A second man, seeing the first run, instinctively copied him, a third and fourth, inspired by two men running, followed suit, with the cumulative effect that soon all were running, and the habit, once started, remained as a fixed tradition. At any rate, it did so while the Welsh Highland Railway continued to bring them down. I have often wondered if the first man had jumped over an imaginary obstacle, all the others would have done likewise.

At the Croesor Quarry we had twenty-two separate groups of slate makers, each with a place on the Quarry bank on which to stack his slates. The rows of slates would start square with the wall against which they were pitched, but after a short while they would begin to turn to the left, and if the row was long enough, I have known it to end up as much as a full right angle to the direction started, thereby impeding the straightforward advance of the pile next to it on the left. This applied universally to every pile on the Quarry bank, causing not only unsightliness but inconvenience, and stranger still, slatemakers coming from any other quarry to Croesor did the same kind of thing, so I conclude it was just another example of unreasoning imitation.

Another example of lack of reasoning was to be found in connection with saw tables. The tables did not advance at a continuous and regular pace, but in little jumps of about a quarter of an inch at a time, but this was due to the fact that it took more energy to start the table from a stationary position than it did to keep it going when it was once started. The tables were pulled forward by a long chain feed operated from a revolving drum, and the long chain having a great deal of elasticity permitted extension to be built up enough to start the table from its stationary position, and the table once started then moved forward a short distance until the tension built up in the chain had exhausted itself.

The tables almost universally in use were made at the Union Foundry, Portmadoc, and the feed drum was driven by a ratchet and pawl arrangement, the ratchet wheel being driven by two pawls actuated by eccentrics on the saw spindle. The men always called these pawls "paddles". The action of these pawls was slightly intermittent and so far as I can recollect every slate maker I have ever met attributed the jerky feed to the intermittent action of the pawls. A very superficial observation of what was taking place would have shown anyone that the intermittency of the action of the ratchet wheel and pawls had nothing whatever to do with the jerky feed of the table, because they did not in any way correspond in point of time, nor in their frequency, yet someone, perhaps forty or fifty years ago started this fallacy and almost everyone else adopted it without question.

This is the more surprising where slate makers are concerned as the same men show marvellous skill when splitting slates, enabling them to put the splitting chisel in the exact middle of a piece of slate, which is only one third of an inch thick, apparently without looking at it or making any effort to obtain an equal thickness of slate on each side of their splitting chisel.

I have dwelt an unjustifiable length of time with this instinct of imitation, but I can hardly stress too much its effect in our lives. For good or ill I was born with the faculty of imitation almost lacking. I could never imitate anything successfully; for instance, I could never learn to dance. My teacher would say: "Now look at my feet and do as I do". He might as well have told me to stand on my head. Before I could get the right steps I would have to make plans, showing the various positions of the feet and the sequences of the steps, but as this was impracticable in a dancing class, dancing is not one of my accomplishments.

This lack of the easy way of learning to do things, and of acquiring ideas, ie by imitation, imposed upon me the necessity of getting my results by other means, and developed in me a hypercritical type of mind which made me sceptical of authority and disinclined to follow the routine followed by other people.

I divided statements seen in print or communicated to me orally into three classes—right, wrong, and not proven, and I think most of them went into the third category. It is really astonishing that things that are regarded as axiomatic are generally wrong, and even the formulae found in text books collected at random by the compiler are often as not incorrect and misleading.

Having this type of mind I am afraid it makes me a very disagreeable person. I see fault where others see none, and a fault-finder, however much he may be justified in his faultfinding and criticisms, is regarded as a nuisance and a disturber of the peace. However, it has its compensations, for when one gets right down to first principles one finds out a lot of things that the more easy-going imitator never discovers.

As a boy, I was very shy, and although I had an innate love of music, I did not join the choir until my voice had changed and I was able to sing bass. From the outset I found that the majority of the choir took their cue from the singer sitting next to them, and sang the same notes as they did. A soprano sitting next to a tenor would sing tenor and would be quite unaware that she was not singing soprano. The tonic sol-fa copies were of little use except to give the words, and certainly not one in four had ever taken the trouble to learn the time symbols. From the outset I not only learnt the time symbols but everything else connected with the reading of music, and when the conductor left Croesor I was astounded when, at 19 years of age, I found myself elected as conductor of the choir there. I immediately pitched into the job of training them in what I considered to be more scientific methods of learning to sing, and the great success of the choir in choral competitions around the country justified my methods.

About 60 years ago music in North Wales had reached a standard far ahead of what it possesses today, and the great musical event of the year was the Harlech Festival, which was attended by the principal choirs of the surrounding counties. These choirs included in their number such fine societies as Portmadoc Choral Society, Aberystwyth Choral Society, numbering each over 200 voices, and regularly giving grand orchestral concerts on their own account.

In very recent months, Bob Owen MA, the famous Welsh literateur was rummaging through old periodicals, of which he has a wonderful store, when he discovered among them a "Cerddor Cymreig" dated 1883. The reporting on the Festival had been dedicated to the famous Welsh composer and critic "Alaw Ddu", and in his report he said that the neatest rendering given by any choir during the Festival had been given by the Croesor Choir. The Croesor Choir was drawn from a village containing less than 200 inhabitants, numbered only 55 voices, and its conductor was Moses Kellow, aged 21, who had the temerity to bring his little choir to a festival where hitherto only large choirs of high reputation had been admitted.

In explanation of the comparison made by "Alaw Ddu" it may be mentioned that at the morning and afternoon meetings each choir sang its own chosen pieces, and at the evening meeting the massed choirs performed oratorios like the "Messiah" or "The Creation" with full orchestral accompaniment, and professional singers taking the parts appropriate to them.

With prize money won at competitions the choir purchased a fine American organ which served for its own accompaniment, and also served for the use of the chapel. For a considerable number of years I acted as organist.

Before I was twenty years of age I was elected a member of the School Board of the combined parishes of Llanfrothen and Llanfihangel-y-traethau on which I served for six years.

Regarding my scholastic and educational experience, I made my greatest progress in ordinary education at Davies' Grammar School, and under a private tutor. I was trained in engineering and geology by my father. Before I was twenty-one years of age, I undertook the management of the Park Quarry.

Parc and Croesor Quarry Office (Cathy Woodhead Collection)

3. PARK QUARRY AND SOME NEW DESIGNS

At this Quarry, while the bedding and cleavage had the usual characteristics of what prevailed generally in the district the plane of pillaring more nearly approached the horizontal than the vertical.

This decided me to reverse the ordinary method of developing a quarry, and instead of working the upper galleries first and afterwards the lower galleries, I reversed the order, and as I was able to remove the rock in the plane of pillaring, I could work gradually upwards leaving the rubbish to fill up the space caused by the removal of the rock.

This method not only facilitated the extraction of the slate, but rendered falls of roof or collapse of pillars impossible. It was a system upon which the Park Quarry was worked for the whole of the thirty-three years during which it was worked. Working was suspended in the middle of the Great War (1914-18), as there was then no market for its produce.

I gave evidence before a Royal Commission on this method of developing a quarry, and was asked to make suggestions as to its applicability in whole or in part to the Festiniog group of quarries, where disastrous falls had taken place, and were even more likely to take place in the future as the chambers were extended in depth, and the over-burden on the pillars increased.

The peculiarity of the direction of the plane of pillaring at the Park Quarry had caused it to nearly coincide with the direction of the ribbons (*former bedding planes*) in the manufactured slate, and if the ribbons were cut out, as of course all the large ones had to be, it resulted in pieces of slate having considerable length and little width. As these pieces were perfectly straight, having no ribbons across them to bend them, the rock was peculiarly adapted for slate ridging. I therefore concentrated on this class of product, and made such improvements in methods of manufacture that I was not only able to double the output per man, as compared with that obtained by other makers, but I was able to obtain a very fine finish, with the result that I was supplying about three-quarters of all the slate ridging used in this country, and was also supplying extensively to the Continent.

Slate ridging consisting of the ordinary rolls and wings made a very neat and appropriate finish to a slate roof, but it was not sufficiently decorative to suit all tastes. I therefore designed and patterned it by interlocking slate ridging, whereby the two sides could be interlocked and an ornamental crest cut to any desired pattern inserted as a key, converting the three pieces into a firm and solid whole, which could be laid on the ridge of the roof in cement mortar, like a tile, and requiring no screws or nails to keep it in position. (*See the advertisement at the back of this book Ed.*)

As both in material and labour it was much more expensive to produce than ordinary slate ridging, it necessarily only found a limited market, but a good many examples of its use are to be found in the Portmadoc neighbourhood. I think it was first put on a house in Portmadoc Road, Criccieth. In Portmadoc itself it was used on the exceedingly long slate shed of the Oakeley No.3 Wharf, on Oakeley Cottage near the Festiniog Railway Station, on the shed on Greaves Wharf, on the Wesleyan Chapel in Chapel Street, and on the house of Mr Richard Davies, Tu-hwnt-i'r-bwlch. It may also be seen on the Council School at Talsarnau.

I have no doubt it has impressed a good many people, but only a comparatively small area of slates used in roofing was exposed, and it was this exposed part that represented their covering. I was aware that in Germany slates were laid diagonally with greater economy of material, but these required to be made a particular shape not adapted to the Welsh product. I set myself to devise a system for obtaining the economy of diagonal slating from ordinary stock sizes of slates. A 24x14 laid with 3" lap at the side would expose to the world 231 square inches—an increase of over 50%.

An important firm of London slaters took it up on royalty and a goodly number of houses in London are roofed on this system. I found, however, that my patent was being extensively infringed, and rather than fight expensive patent actions I preferred to drop the whole thing.

Under the scheme of development I had adopted, it was necessary to sink a deep main shaft as a preliminary to opening chambers, and I sank a shaft 14ft square to a depth of five galleries below the main or office floor. This shaft was sunk with the aid of compressed air drills which, however, showed little advantage over hand drilling. The compressed air was afterwards used for pumping the water out of the quarry, and for operating compressed air winches for loading the blocks. The winding from the shaft was done by an 80hp Rigg rotary engine, operating from a head of water of 475ft, thus giving a pressure of water at the head of the shaft of 206lbs per square inch. I conceived the idea of using this head of water for unwatering the quarry, in substitution for the very inefficient compressed air, through the medium of an injector.

There was no such thing as a hydraulic injector on the market, but I knew that steam injectors created a high velocity jet of water in the centre of the tube, and at the surrounding space containing the water to be pumped. The eventual resultant velocity of the water in the injector tube corresponding to the head against which it was desired to deliver water, it appeared to me that provided I could get a jet of water in the centre of my injector tube, of considerable velocity, I could by properly proportioning the motive power with the water to be pumped, get a result analogous to that of a steam injector. But every injector maker in this country, and those in America to whom I applied, flouted the idea, saying that without steam I should get no success. Unconvinced, I let the matter remain in abeyance for twelve months, when I was fortunate in meeting the head of a continental firm. He was the first man who ever gave me any encouragement, and said that he had once heard of a laboratory experiment whereby water had been raised through a small pipe to a height of twelve feet on the principle I had suggested.

My shaft was 225ft deep vertically, and the volume I wanted to lift was large, which was a somewhat different proposition to the laboratory experiment referred to above.

To cut a long story short, I got this firm to undertake to make this special injector of mine, and it was duly installed in the shaft. It was, from the moment of opening the pressure valve, a complete and unqualified success. An absolute river of water was raised from the shaft, without involving the use of any moving mechanism. All that was required to be done was to open the pressure valve to start it, and close it when all the water was out. In the meantime, it looked after itself.

Later on, I designed a hydraulic injector on the same principle for the Rhosydd Company, but under less favourable conditions than those prevailing at the Park Quarry, the pressure head being less, and the lift greater. Nevertheless, its success was absolute, and the problem of unwatering Rhosydd Quarry was solved down to the depth the workings had then reached.

My next experiment struck out an entirely new line. I had often seen a beautiful fire in a grate which required replenishing with fuel, buried by shovelfuls of black coal which largely nullified its utility as a source of heat, giving a dull and uninviting appearance, and incidentally sending out clouds of thick smoke through the chimney.

I designed a grate of nearly cubical shape, with a hinged grid at the top and bottom, which could be secured by a spring catch. This box could be rotated on a horizontal axis, and the procedure when fresh coal was required was first to clear the fire of its ashes, the fill the grate to the top with coal, close the lid at the top, and finally turn the grate upside down, and opening what had been the bottom lid.

There was also an oscillating movement which brought the firebars at the back of the grate against a firebrick surface which would throw back the heat. As these became oxidized a false front, mounted upon hinges at the side was desirable to cover them, so as to maintain an attractive appearance at the front of the grate.

The rotation of the grate brought the incandescent fuel to the top, leaving the unsightly black coal covered up at the bottom. The black smoke and gases arising from the coal at once became ignited as they passed through the incandescent fuel which overlay it, and the whole fire showed flaming gases all over its top, giving a pleasant as well as useful fire for all purposes.

An inspection of the chimney outside the house showed not one vestige of the black smoke that had hitherto been coming out, and pollution of the atmosphere and surrounding buildings was thereby avoided. The grate was simple, and a complete success for what it was intended to accomplish. I only made it for my own use, and I suggest to Major Gwilym Lloyd George, the Fuel Controller, that it may be worth his attention. There is no doubt that besides other advantages, the economy in fuel was considerable.

In January, 1892, I married Nell, the daughter of Thomas and Elizabeth Williams, of Bryn, Croesor. Mr Williams had been the Manager of Croesor Quarry for a very long period of years, and both before and after I married his daughter, I had become fairly intimate with the conditions prevailing in that Quarry. Two or three years afterwards, Mr Williams died, and I took over the management of the Quarry.

Bryn, the Kellows' house just beyond Croesor (Cathy Woodhead Collection)

4. CROESOR QUARRY AND ELECTRIC POWER

The Quarry had not been worked for seventeen years, but from the outset I had planned to have it re-opened. Early in 1896 I arranged the amalgamation of the Park and Croesor Quarries. At the Park Quarry we had nearly every slate merchant in the kingdom as a buyer of slate ridging, which was sent off with slates from other quarries. As there was a bed of Old Vein slate 40 yards thick at Croesor, I saw no reason why we should not send slates of our own manufacture with our ridging.

During the period the Quarry had not been at work, the galleries below the main level were full of water. They contained twelve chambers and there were also two large sinks each containing about 4½ million gallons of water. The general impression in the countryside was that the only way of un-watering it was to drive another tunnel about 400 yards long lower down on the hillside.

A very eminent firm of consulting engineers in Westminster had been asked to advise in the matter some time previously, and had recommended pumping it out with pulsometer pumps, which they reckoned could be done in six months. I had other ideas.

The rails on the incline to the lower galleries had not been removed, when the Quarry ceased working. I made a stage which could be slid down on these rails as the water was pumped out, and mounted on it a powerful centrifugal pump. I lifted the water at the rate of 340,000 gallons an hour, and emptied the Quarry in a week! The quantity of water flowing through the main tunnel, which was nine feet wide, was so great that it would have been easily possible to go in or go out from the Quarry in a boat.

Very shortly after starting work at the Croesor Quarry the disadvantage of letting the bargains to rockmen who were in partnership with slate-makers in the mill was forcibly brought home to me, as it had previously been brought home to me at the Park Quarry. *(Note: A bargain was a group of, say, four men who were allocated a chamber where the two rockmen extracted slate and send it to the mill for the two quarrymen to saw, split, and dress the slates; they would bargain with the quarry steward on the price at the beginning of the month and they were paid on the last Friday of the month collectively for the number of slates produced.)* The flow of blocks from individual bargains was exceedingly variable. Interruptions occurred while free bottoms and free sides were being cut out. These interruptions often lasted for several weeks.

In the meantime, the slate-makers associated with that bargain had nothing, or next to nothing, to do, but after the preparatory work referred to had been completed, rock could be taken down and sent out much faster than the slate-makers normally provided to cope with it could use up, and the rockmen inside the Quarry were unable to work to their full capacity owing to the glut of unworked blocks.

It often occurred that the exact reverse of the conditions enumerated was occurring in an adjoining bargain, but the conditions of the contract precluded the use of the surplus of one bargain to make up the deficiency of the other.

I decided to separate the contracts of the rockmen and the slate-makers, and pool the supply of blocks from all bargains and thereby avoiding the compulsory idling both above and below ground, which resulted from the old system.

Needless to say, so drastic an innovation met with great opposition from the workmen who wanted to be allowed to work in a manner similar to that worked by the men in other quarries.

However, in spite of this opposition I carried out my plan, and the first month it was tried the output increased by 25%, and incidentally the average wages of the workmen substantially increased. I worked on this plan during the whole of the period that Croesor Quarry was in operation, and had the satisfaction of having my example followed by other important Quarries in North Wales.

In 1901, I prepared a scheme for the electrification of the Croesor Quarry, and proposed to derive the power from the Cwm Foel Valley, which was the source of the Croesor river, and which was 1460ft above sea level.

By constructing a reservoir in this valley, and laying pipes and building a hydro-electric generating station at the foot of the Croesor Valley. I was able to utilize a head of 860ft, which I think was the greatest that had up to that time been developed in the United Kingdom.

At that time electrification for power work meant direct current at 550 volts, an undesirably high voltage for distribution underground. Unless, however, a high voltage was adopted, the cost of cables for transmission and distribution became prohibitively high, and standard practice had prescribed 10% loss as the most desirable compromise.

I found the cables, however, to be too costly on the basis of 10% loss, and had eventually adopted 13% as my transmission and distribution losses. But I was very dissatisfied at this high figure and held up the scheme for a while for further investigation.

During this period I came across a reference to a new system of electrification which had been discovered, and this was the use of multi-phase, and in particular three-phase alternating current. Under this system high voltage could be used for transmission, and stepped down by stationary transformers to any low and convenient voltage desired for distribution. The three phases gave a rotary motion for starting analogous to that of a three-crank engine, and last but by no means least, the commutator and brushes which caused nine-tenths of the trouble in direct current machines was eliminated.

I was very impressed, but the condemnation of 3-phase by British electrical manufacturers was almost universal, and to adopt 3-phase meant not only going dead against British electrical authority, but it meant also that I could not get it made in this country. I felt convinced that it was the best system for me to adopt for Croesor, and to crystallise my own ideas I drew a full and detailed specification of the 3-phase plant in 1902.

I did not issue this specification, but took it with me and made an extensive tour of Europe, visiting most of the works of the electrical manufacturers. I found them much more favourably inclined to the 3-phase system, and they confirmed my view that for power work it was preferable to the direct current system. Its adoption, however, entailed a certain amount of risk, as nothing had really been standardised.

Alternators were being made with rotary armatures, rotary field magnets, and stationary armatures and inductor types. I selected the rotary field type with stationary armature, and the machine supplied would be regarded as an up-to-date machine today. It is still in use, and has needed but one slight repair all these years.

Periodicities were all over the place, and motors were being made of a variety of types. Moreover, no electrical rules existed in this country covering the new system, and it was a leap in the dark to install an electrical system which might not conform to Government regulations later on. There was also the difficulty of cables, cable-makers had not learnt how to make cables able to stand the high voltages utilized in the 3-phase system. The plant was manufactured by Kolben & Co, of Prague, and was delivered in due course. The alternator was 250 kva, directly driven by a 375hp impulse turbine. In order to obtain the best possible voltage regulation, I drove the exciters separately by a 25hp impulse turbine. The duplication of the exciter was to provide against breakdown of the generating plant by the failure of its most delicate part, and I was able to use the second exciter for lighting and heating my house.

Transmission was nominally of 3000 volts, but I kept it a little lower to avoid coming under the very stringent rules which were then pending for extra-high tension at the Quarry. I transformed down to 220 volts between phases for all purposes. I adopted group driving for the mills, these consisting of 10hp squirrel-cage motors geared directly to sections of the lineshafting.

I outraged orthodox electrical opinion by switching these straight on to the line, without attempting to reduce the heavy starting current, but I obtained thereby a starting torque which would start every saw table from rest, even though the teeth were in contact with the slate blocks. The troublesome alternative would have been to throw the tables out of gear and wind them back by hand, so that the saws were not in contact with the blocks, and the motor could thereby start unloaded.

Haulage in the main tunnel was done by a 30hp electric mining locomotive, which was the first electric mining locomotive that had been used in this country. Speeds varying from one to eight miles per hour were provided for, and a tractive effort was designed for a train of 12 loaded wagons, weighing about 30 tons. I have seen twice this number successfully hauled.

For distribution below ground I arranged a transforming sub-station. A 90 bhp 3-phase motor with wound rotor was used for the main haulage incline, and I provided a reversible controller whereby if a load was started up the incline and found to be back-heavy, it could be wound back again and adjusted.

The bargains were provided with reversible winches, each driven by a 10hp squirrel-cage motor. These were geared to give a three ton pull on a rope for loading blocks and an alternative high speed for driving them along the floor. This rope pull directly from the barrel dispensed with the use of cumbersome pulley blocks, and the reversibility of the winch enabled the wire rope to be wound out after the lifting or hauling had been completed.

Needless to say, all the shunting siding, inclines, tunnels etc., were electrically lighted, but for the first time in underground slate quarrying the chambers were lighted electrically, each chamber had two arc-lamps, each of 2,000 candle-power, suspended overhead, and the whole of the chamber was illuminated as though by daylight.

Means were taken to raise the arc-lamps out of the danger line during blasting. Later on, it was possible to obtain ½ watt, metallic-filament lamps of 1500 watts, thus giving about 3000 candle-power, and these were found much less troublesome as well as less expensive to use than the arc-lamps which they replaced.

Such was the electrical experiment, but I also carried out experiments with the impulse wheels, the designs of the buckets and the nozzles of which I considered to be defective. I subsequently described my improvements in turbines in a paper read before the Institution of Civil Engineers. Twenty-five years later I had the satisfaction of seeing my buckets and nozzles, and the arguments I had brought forward for their adoption, quoted in a large two-volume publication by a well-known engineer as representing the latest practice.

Up to the time I put in the electrical plant at Croesor Quarry, drainage when operated by electricity was done by three-throw ram pumps, mounted upon a massive bed-plate carrying at least two trains of speed-reducing gear. The highest speed shaft carried a belt pulley driven by a wide belt from a 500 volt electric motor. The whole arrangement was costly, occupied a great deal of space, and the quantity of water lifted was very small, It appeared to me that the marriage of a slow moving reciprocating pump with a high speed rotary motor was an illogical combination, and that the better way would be to directly drive a high speed rotary pump by a high speed rotary electric motor, but the only type (worth mentioning) conforming to these conditions was the centrifugal pump, which, however, was only adopted for low lifts.

In the technical press of the period there were illustrations of pumps that were being made by a well-known firm, in which the delivery pipe of one pump was connected by outside piping to the eye or suction of a second pump placed in line with it. This gave an additional impulse to the water and enabled it to be lifted to a greater height than would have been possible from a single pump. I approached this firm and asked them what efficiency they were prepared to guarantee for this combination, and was told 40%. I replied that I would not put in a pump having so low an efficiency.

Later on, during my travels on the continent, I came across a pump constructed on the modern series-centrifugal pump model, with all the stages assembled together in a neat cylindrical casing. What was more to the point was that, on calculating the litres it was claimed that the pump would raise against a specified head and from the stated watt input, I found that the efficiency came out to about 80%.

Very shortly afterwards, I visited the makers of this pump, stated my conditions to them, and obtained a guarantee of not less that 75%. The pump that was eventually installed, which lifted 400 gallons of water a minute, actually gave 80% efficiency, and is still in use. So far as I have been able to ascertain, this pump was the first series-centrifugal pump of modern design which had been put in industrial operation in this country.

About 1907 I was informed that the Rees Roturbo Co were intending to make a new issue of shares and had applied to the Industrial and General Trust to underwrite it. On behalf of the latter company, I was asked to report on the pump which the Rees Roturbo Co had been formed to manufacture.

From the papers sent to me it appeared that the makers of the pump put forward a very important claim, i.e., that they had discovered a new source of power which would reduce the power required to work their pumps. The impellers of the pump consisted of a drum from which, when rotated in a clockwise direction, the water was discharged through a restricted orifice in an anti-clockwise direction, and the reactive effort of this on the impeller lessened the power required to drive the pump, and the claim was that an increase in efficiency was thereby gained.

This claim was undoubtedly supported by the fact that a pump constructed in this way took less power to drive it than one made in the ordinary way. So far as I am aware this claim was not disputed in the trade generally and the company had built up a substantial trade in their pumps.

On the principle, however, that you can never get something for nothing, I did not accept the view that there could be any gain in efficiency from this source. The water in the drum of the impeller had to be put in motion, and this absorbed energy. To re-direct the water opposite to that in which the impeller was travelling had, in my view, the effect of taking out velocity from water that had been put into a state of velocity by the expenditure of power and therefore must of necessity lessen the energy remaining in the water which was available for doing useful work. Although I put forward this view in my report with absolute confidence, my report was not hostile to the pump, and I think the Rees Roturbo Co got the underwriting they required.

Shortly afterwards I visited the Exhibition at the White City, and opposite the main entrance I was confronted by a gigantic cascade of water, making the exhibit to which it belonged one of the most impressive in the Exhibition. On investigation I found that this water was being delivered by a Rees Roturbo pump. I got into immediate contact with those in charge, to whom I believe my views had already become known, and they took great pains to try to prove to me that their claim to have discovered a new source of power was correct. They even went so far as to stop the cascade for 1½ hours, while they took the pump to pieces to show me its construction. I was not convinced, and they begged me to go to their works at Wolverhampton, to discuss further the matter with a technical staff there. I did so, and first went through the matter carefully with their chief designer. I still held firmly to my view, and it finally came about that I had not only the chief designer in my company, but the whole Board of Directors as well.

I said: "I am in want of a centrifugal pump to give 90 gallons a minute at 750 lbs pressure per square inch, and can buy such a pump from another maker with a guaranteed efficiency of 60%. If you can gain power, as you claim, you ought to be able to do better than this. What will you guarantee?" The answer was: "We will guarantee 65%, and not charge for the pump if we do not give it." I replied that I did not think they could give it, and that I did not wish them to give me a pump for nothing. The reply was: "We would like to have the opportunity of proving to you that our claim is justified".

I accepted their sporting offer, and the pump was made and supplied in due course, but as I had expected, the guaranteed efficiency of 65% was not fulfilled. I had the means of making an accurate test, as I had a watt-meter on the motor circuit, and a Venturi water-meter on the delivery side of the pump.

I invited the Rees Roturbo Co to send someone to verify or otherwise check my conclusions regarding the non-fulfilment of the guarantee. Two engineers from the Company came to the Quarry where the pump was at work, agreed the accuracy of my test, and like the gentlemen they were, the Rees Roturbo Co made no charge for their pump.

I would, however, have preferred paying for the pump and getting the guaranteed efficiency, to getting the pump at a lower efficiency for nothing. It was not my business to publish to the world what had taken place, but twelve months later one or two papers began to appear in the technical press disputing the Rees Roturbo Co's claim to have discovered a new source of power, and after a good deal of correspondence in the press, I think the general consensus of opinion was that their claim was not a tenable one.

In connection with the electrical plant at Croesor Quarry, the manufacturing firm in Prague quoted a very high price, but gave guarantees of efficiency, power factor, starting torque, high inherent voltage regulation in the alternator, and otherwise put forward claims of a very exceptional order. Personally I did not believe they could ever be attained, but the firm assured me they could be, and would be upheld. I agreed the price, subject to the upholding of the guarantees being the essence of the contract, and this condition was accepted.

It was specified in the contract that I should go to Prague to test the plant before it left their works, and this I did exhaustively, occupying a very busy four days in doing so. I found that, although a mechanically and electrically good plant had been made, the extraordinary high guarantees had not been reached.

When I first arrived at Prague I was told that the plant had been tested and that all the guaranteed figures had been attained. It was suggested that I should accept their figures, instead of going to the trouble of testing the plant myself, but, consistently with my habit of taking nothing for granted, I tested the plant with the results already stated. When they found that I knew the facts, they showed me the results of their own tests, and in no case did they vary from what I had taken by more than ½%.

I allowed the plant to come forward, as we still had the right of rejection if it proved unsatisfactory in actual operation. I deplore what followed.

Erection was to have been completed within six months, but the turbine makers were not ready with their part of the plant. The manufacturers, through their agents in London, gave us notice that they would claim substantial damages for the delay, and without naming the amount of these damages, demanded an arbitration.

I did my utmost to come to terms with them, offering to waive the non-fulfilment of guarantees, and even offering a cash sum in addition.

This was rejected, and an arbitrator was nominated by them. This man being the man from whom they purchased their cables! Finally, we both agreed on Mr Charles Peace Parks, the Chief Engineer of the County of London Electric Supply Co, as the sole umpire, and a good many hearings took place at the Great Eastern Hotel, London, and one of them at Croesor.

When the arbitrator came there to inspect the plant I was fully prepared. For the first time in my life I acted as counsel in an important or critical case in addition to being the only technical witness, and I had opposed to me technical experts in 3-phase plant who had been trained abroad under the pioneer of 3-phase electricity in Europe.

I appeared to carry the arbitrator with me throughout the proceedings. I got every point I was fighting for, incidentally got the right of rejection of most of the motors, which I afterwards bought back at half price, and my opponents came off more than a thousand pounds worse off than I had offered them at the outset.

In 1908 I was requested by the Council of the Institution of Civil Engineers to read a paper on The Application of Hydro-Electric Power to Slate-Mining, and had the honour of being entertained at dinner by the Council at what was then the Westminster Palace Hotel, afterwards known as Abbey House, in which the Oakeley Slate Quarry Company have had and have a fine suite of offices.

As an indication of the doubt that still lingered in the minds of the most distinguished engineers regarding the general adaptability of the 3-phase system to industrial purposes, I was warned before I read my paper by Mr Tutsbury, the Secretary of the Institution that I would have great difficulty in persuading its members of the possibility of using 3-phase current for haulage and winding, and he advised me to come well prepared with facts and arguments on this point. As a matter of fact I had been using it for this purpose for a number of years.

As it was thought that the discussion on my paper would be directed mainly on this point, it was decided to read another paper on electric winding in collieries by Messrs Preece before taking the discussion on either, as it was thought probable that they would overlap.

Though I went fully prepared to encounter criticism on the suitability of 3-phase current for haulage and winding, I found none, and there was no question of the two papers overlapping.

In my introductory remarks I had been rather apologetic at presenting the Institution with the details of an installation totalling only 400hp, whereas they were accustomed to engineering installations of vastly greater magnitude.

On the other hand, I said that it had the unique character of being very comprehensive in its scope, beginning at the rain drops and involving problems of the storage of water, conveyance through pipe lines, the generation of mechanical power by turbines, of 3-phase current by alternators, transmission at extra-high voltage, transformation, distribution at a secondary and lower voltage, and its ultimate applicability to the driving of motors and to lighting.

In fact, it covered the whole range from the rain drops to the motor shaft or the lamps. Later on, in the supper room, I overheard two members discussing my paper, and one remarked to the other: "Its scope was as wide as that of the Institution itself".

However, the paper was exceedingly well received, and later on I was awarded a Telford Premium, which took the form of a very large induction coil capable of giving up to 750,000 volts, Caldwell interrupter and a goodly collection of mineral and other vacuum tubes for the study of electrical phenomena in high vacuum.

The box containing the induction coil has a silver plate upon it, stating that it was presented to me by the Institution of Civil Engineers. The Times newspaper published an article on my paper, and I was astounded at the publicity obtained all over the world through the medium of the minutes of the Institution of Civil Engineers, and of the Times notice.

In fact, shortly after I had read the paper in 1908, when I think my reputation stood at a fairly high level, I was approached by a Member of the Institution of Civil Engineers to know whether I would be prepared to undertake an investigation into the Lots Road Electrical generating station, belonging to the London County Council. This was at the time the largest electrical generating plant in this country, and I believe in all Europe. The overall efficiency of the station was not up to expectations and it was desired to find out at what stage or stages the fault was occurring.

This would have meant taking the scientific test of the calorific value of the fuel, the efficiency of the boilers, of the engines, and of the electric generating plant separately and independently. I was told that it would be a whole time job for six months. If I had accepted the proposal, it would have meant taking up my residence in London, and throwing up all my other activities. I was informed that the London County Council had asked the Institution of Civil Engineers to recommend an Engineer to carry out this investigation, but I cannot vouch for the accuracy of this statement, nor whether the man who approached me came as a result of this or not.

As for some time after reading my paper I was overwhelmed by correspondence from places as far apart as South America, Australia and Europe, concurrently with my activities in the hydro-electric field, and was in whatever spare time I could muster devoting my mental energies in experimenting with rock-drills, I was compelled to reject this Lots Road proposal.

The site of the Croesor quarrying works with the 'road' down to Croesor just visible in the middle of the photograph and Blaencwm power station on the right in the valley bottom (Cathy Woodhead Collection)

For a review of Kellow's contribution on this topic see Tim Oulton's 'Electricity in the North Wales Slate Industry', in Mining Technology: Technical Innovation in the Extractive Industries. Proceedings of the NAMHO Conference 2014, ed. D.J. Linton, Welsh Mines and Mining 4 (Welsh Mines Society, 2015), pp. 147–158. Ed

5. THE KELLOW DRILL

In my early boyhood, even before I was nine years of age, I had become aware of the inefficiency of compressed air as a means of transmitting power, and my later experience strongly confirmed that view. I therefore decided to use high water-pressure instead of compressed air for the reason that the power-weight ratio was higher in a turbine than in any other form of motor ... hence portability.

Water could be used for washing the debris out of a hole, cooling the bits, and by means of an hydraulic cylinder, giving the ideal form of feed. In my initial experiments I had some strange surprises. A bit directed against the rock, instead of penetrating in the direction required, would go in almost any other direction conceivable, and would then jam. Moreover, I found that spiral bits would not clear the debris for a greater depth than one foot. Eventually, I solved the problem not only of getting the hole to go straight, but of getting exceedingly rapid penetration.

Mr Treglown, the Chairman of the Ingersoll Rand Co in Europe, asked to see my drill at work, and came to the Park Quarry where it was being worked at a pressure of 206 lbs per square inch for the purpose. A hole was drilled in one and a half minutes, and his remark was: "We haven't a drill that could have done that in twenty minutes. Take your drill to the United States, and fix up with our people over there. We will defray all the expenses of the journey and the demonstration".

I accordingly went to America, taking with me the drill he had seen at the Park Quarry, gave the demonstration, and brought back an offer that we should manufacture the drill in this country and that the Ingersoll Rand Company would, through their vast selling organisation, sell the drill throughout the world, and that the profit should be equally divided.

I have never ceased to deplore that this offer was not accepted, as at this early stage of the development of the drill it would place at our disposal the vast resources of the world-renowned Company.

I am not a believer in Luck, otherwise I would say that in the efforts to exploit the drill we were on several occasions pursued by downright bad luck, the one mentioned above being an example. I prefer to call it the result of fortuitous circumstances, which in their result proved unfavourable to the Syndicate.

The drill had undoubtedly established its success from a utility point of view. Up to the time it was introduced, practically all drilling in slate quarries had been done by hand at the Croesor Quarry, and this was typical of the Festiniog district generally. It took a rockman a whole day of 10 hours to cut a hole 7ft 6ins deep with a jumper. I do not think on the average that more than 6ft a day was actually drilled per man in the district. With a Kellow drill a hole 7ft 6ins could and was being drilled from start to finish in 1½ minutes.

I once had a rather amusing experience with a well-known engineer, who had come to see the drill at work in Croesor Quarry. I told him we were going to drill a hole 7ft 6ins deep, and as he did not expect that this could be done in less than about an hour, he prepared a comfortable seat for himself while he waited for the operation to be completed. He settled himself down, and the drill was started. It worked almost noiselessly, and nothing could be seen happening except a little trickle of water coming through the hole.

At the end of one and a half minutes the hole was completed and the drill stopped. The Mining Engineer jumped to his feet and exclaimed: "What has happened? Has anything gone wrong?" "No," I replied, "but we have finished the hole." He simply would not believe it, so a stiff wire was produced and he himself pushed it down to the bottom of the hole, and measured its length incredulously.

"I do not believe it," he said. "You had a hole drilled there before, and filled it with soap".

I am sure the good man left the quarry half-suspecting that a trick had been played on him.

I was paying from 6d to 9d per linear foot for drilling at Croesor Quarry, but after the drill was made available, one man did the drilling of the whole quarry, consisting of 14 bargains widely distributed over four floors, and the price paid was ½d per foot, including transport from place to place.

I may add that about this time I was approached by the General Manager of one of the most important mines in the world, the name of which I am not at liberty to disclose, at which untopping was done in galleries 40ft deep. The practice was to put horizontal holes 9ft deep at the bottom of the working face, then drill vertical holes 40ft deep, 9ft back from the face of the rock, and blast these holes simultaneously by electricity, thereby bringing down a fairly large quantity of rock which was removed by steam driven navvies.

The Manager was desirous of increasing the 9ft horizontal holes up to 20ft, thus lessening the number of vertical holes by more than one half. He consulted me professionally with regard to the possibility of doing this, not knowing that I was the inventor of a rock drill.

I asserted that it was not only possible, but that I could send him a drill that could do the work. He advised me to be cautious in making this promise, as he said that the largest rockdrill company in the world, using cylinders up to 5½ inches diameter had hitherto failed to drill more than 15ft horizontally, because the debris got under the drill point, and caused it to turn upwards and jam, notwithstanding the large diameter which had been provided in the rock-drill cylinder solely for the purpose of pulling the drill back.

I assured him this difficulty would not arise in the case of my drill because the debris was washed out by high pressure water A drill was accordingly sent out, and the first hole drilled was 27ft deep by 3" bore, this bore being not only

3" at the top but also 3" at the bottom, instead of being 3" at the top diminishing to about 1½" at the bottom, as would have been the case with the ordinary type of drill.

The hole was completed from start to finish in 37 minutes, including the time taken to put on extension rods and change the cutters. Apart from the fact that hole had been drilled to a depth of 27ft, the rate of boring was at least 10 times that of any other drill on the market.

With such records as these, it was considered that the time was ripe to float a company for it, and a promoting firm in London undertook to form a company with a capital of £120,000, and at their offices in London I addressed audiences composed of the most important mining and mechanical engineers in London, stock-brokers, etc., of the principles and capabilities of the drill generally, and met with an all-round good reception.

After one of these meetings, two men who had been present remained behind and intimated that they wished to take up the drill for Mexico, where they had extensive mining interests.

Their offer was to form a company for £30,000 and to allot us £15,000 in paid-up shares for the Mexican patent rights. This was considered satisfactory, and was provisionally accepted.

They had hardly left the office when another man who had been present at the meeting, came in and said: "Mr Kellow, I have come to make you an offer".

"Whom do you represent?", I asked, and he smiled.

"Ten millionaires!", he answered briefly, at which remark I concluded that I was having my leg pulled, but to my astonishment he proved his bona fides, showing he did actually represent ten millionaires, heads of ten mining companies, each of which had a capital of £1,000,000. He said that the company of £120,000 which I was contemplating was too small, and that a capital of a million pounds was necessary. They would put up £10,000 in cash immediately, and subject to this being satisfactory and our giving them the world patent rights, they would find the million capital among themselves, and would allot us one-third of the capital for our rights.

I told him that I had virtually fixed up for Mexico, but I was able to arrange almost immediately with these people not to ask for the exclusive Mexican rights, but to join the larger Syndicate.

I had brought the negotiations up to this satisfactory point when I received a wire that our works in Wales had been completely burned down, that every machine and drawing had been destroyed! It took eight months to make fresh drawings with a view to carrying on negotiations from the point where they had been suspended, but on returning to the man who was representing the powerful syndicate, his comment was: "Mr Kellow, you can never re-heat an iron which has once become cold", and I found it impossible to renew the negotiations.

I may here remark, however, that this Company never lost its interest in the drill, and witnessed demonstrations in drilling very hard granite, very abrasive rocks, channelling etc.

I will pass over a very interesting but not financially remunerative history of the drill, but I learnt that while one man may invent a mechanically successful drill, it takes more than a hundred or even a thousand men to finance or market it, and £100,000 as a fighting fund to protect the patents against infringers.

To return briefly to the capabilities of the drill, I may say that I used it as a channeller, and traversed it along a horizontal support; with holes 2½" diameter drilled about 6" apart, the intervening spaces being afterwards bored out with a 3½" diameter drill, work was recently done at the Llechwedd Quarry, Blaenau Festiniog, at almost incredible speeds.

The 2½" holes were bored to a depth of over 7ft in 1¾ minutes, and the 3½" holes were similarly bored to a depth of over 7ft in 2¼ minutes. I cannot but believe that in the future these machines will have a profound and very beneficial influence on the working of slate quarries, and have a revolutionary effect on the methods of quarrying slate, substituting the scientific cutting out of the rock for the unscientific and barbarous method of blasting it out, and shattering it all to pieces.

About 1904 I was appointed by the Festiniog District Slate Quarries Proprietors' Association, to put evidence on its behalf before the Royal Commission on industrial diseases, the evidence to consist chiefly on the amount of dust produced in the quarrying of slate, and in its manufacture in the mills into commercial roofing slates. The chief offender as regards the quarrying of the rock was then the hammer-drill, the dust from the holes being blown out by compressed air passing down through the centre of the drill. As regards manufacture in the mills, I was able to prove that the amount of dust produced by the various manufacturing processes was practically negligible.

The circular saws which would have been most likely to produce dust being run in troughs of water, and the dressing machines by their shearing actions cutting off large fragments produced very little dust. Dust which had accumulated over a long period of years on the tops of walls or other flat surfaces was collected and weighed, and the results in relation to area were given.

I was also able to tell them that an hydraulically driven drill—the Kellow drill was in use in connection with the quarrying of rock at that time—was absolutely dustless.

I gave figures showing the very high efficiency of transmission of power by high pressure water conveyed in pipes for working rock drills or other machinery below ground, and the Commissioners took a good deal of notice of this, and later in the day, having been invited to take lunch with them, the Chairman, Sir Henry Cunningham, said to me: "The great field of usefulness for your drills, Mr

Kellow, will be the collieries. Don't you agree with me, Redmayne?" Sir Richard Redmayne was the Chief Inspector of Mines for the Kingdom, and in his early days had been a colliery manager. He replied: "I quite agree. Water is a very good thing to have about a Colliery!"

(See the advertisement for the Kellow Drill at the back of this book. Ed)

The Kellow Rock Drill working on a near vertical face

6. SAW TABLES AND OTHER INNOVATIONS

I also directed my attention to other machinery used in slate manufacture, and it is remarkable when one comes to consider matters critically, how terribly inefficient machines can be, and yet so long as they do a certain amount of work their continued use is tolerated.

Saw tables used for the cross cutting of slate blocks is one example of this. Blocks for splitting into slates are usually 2½" thick, and although slate is not a hard substance, the usual rate at which it is sawn is only 3" a minute, and it takes about 40 minutes to saw a tableful of about 11ft. The principle cause of this inefficiency is to be found in the shape of the teeth of the circular saws employed. A little reflection will make it obvious that it is the front of the tooth that comes into contact with the slate block. The teeth of all slate saws sharpened by the ordinary filing machines have a negative angle on their front. In other word, they slope backwards.

Now it is obvious the material the tooth is intended to be used against is being pushed towards the saw centre and not away from it, and the tooth therefore is being pushed into the block. The escape of the material this tooth has to remove is rendered extremely difficult, and involves a great deal of friction and unnecessary pressure.

Notwithstanding the fact that the lower part of the saw is travelling in a trough of water, there is little doubt that soon after the tooth of the saw enters the block, its extreme point is very nearly if not quite red hot. It gets into a plastic condition in which the metal is pushed away and the part immediately behind the point of the tooth begins to rub against the block, causing more heat and friction.

Anyone observing a slate saw being removed after it has been working only a short time will notice that at least an eighth of an inch of the tooth has been worn away, and that the tooth has been in a very softened and plastic condition is shown by the fact that the metal has actually been pushed over and backwards on the rear edge of the tooth.

Because the saw filing machine always gives a negative angle on the front of the tooth, I condemned them as being quite unsuitable for the work they are intended to perform. The teeth of any saw should always have a positive or cutting angle, so as to force the material to be removed away from and not towards the block. An example of the ideal form of tooth is to be seen in a timber saw, and if anyone accustomed to the performance of the ordinary slate saw was to see how a timber saw, with its correctly formed teeth, can saw slate, it would be a revelation to him.

Under the adverse conditions that ordinarily prevail when the usual slate saws are in use, the pull upon the table becomes so great that I have often known the chain feed to break under the strain, although its breaking strength was about 1 ton. Most people have some familiarity with the astounding extent of astronomical distances, but it is not everyone that realises the really astonishing which, in the direction of minuteness, indicates the dimension of molecules and atoms, and yet it is with molecules that we are dealing when we attempt to separate particles of slate from the parent block. Could we but visualize what is taking place in the process, the evidence in proof of what I have said would become manifest, and logical deduction as to what is taking place must not be ignored because of the minuteness of the particles that are being dealt with.

The next point to be considered is that in a slate saw of 24 inches diameter there are about 75 teeth, and as the saw revolves at 50 revolutions a minute about 3750 teeth have to pass to advance 3" into the block, which means 1250 or 50th part of an inch for each tooth.

Now it is never economic to attempt to remove material in unnecessarily small particles, and coarseness of cutting is more economic of energy. As a matter of fact, owing to the springy feed due to the chain, unequal length of teeth, and bad setting, and the eccentric way in which the saws are nearly always mounted on the saw spindle, the teeth are never taking their proper and equal share of the work to be done.

The next point is that a great enemy to efficiency is found in vibration, which is capable of absorbing a tremendous amount of energy in an entirely useless form. It is more often found than not that the moving table of the saw-table, which should be supported by rollers, will be in contact with only a few of them, and the unsupported part is in a state of constant vibration.

By attention to all the above points (the teeth of the saw having a proper cutting angle on its front side, teeth of the same length correctly set in a screw press, adjustable rollers to support the table without unduly large interval, screw feed on the moving table instead of chain feed, a large diameter saw shaft to eliminate springiness, a driving pulley with a heavy rim to act as a flywheel, but without increasing the speed of rotation) I was able successfully to increase the rate of cutting from 3" to 3ft a minute, and the thickness of the block sawn from 2½" to 6 and even to 8 inches.

The working was perfectly smooth, and the blocks were not heated nor adversely affected in any way. I used this table for many years for sawing up the long blocks used for slate wings, and it only required about 3 minutes to saw a tableful as against the usual 40 minutes. As an indication of the tremendous waste of power which takes place in an ordinary saw table I was able to obtain the greatly increased result mentioned with the expenditure of only between 1½ and 2 horse power.

I now directed my attention to planing machines. We had a massive planing machine which had been constructed for iron planing with a platten 12ft x 5ft. It ran normally at 12ft a minute in the cutting stroke and 18ft in the return stroke. Its weight was three tons. The accepted view of engineers in general is that while rotary machines can be speeded up to any desired extent, reciprocating machinery, like planing machines, can not be increased to any great extent because of the momentum which has to be imparted to moving parts during acceleration and dissipated during deceleration, and that the chief factor in this was a very heavy platten, which had to be reversed at every stroke.

On investigating this problem more thoroughly, the fact emerged that not only had the heavy platten to be reversed at each stroke, but that all the gearing necessary for driving it had also to be reversed. Now, momentum varies as the square of the velocity, and whereas in some planer gears the ratio of belt speed to platten speed was 80 to 1, every pound weight in the rim of the belt pulley represented as much momentum as 80 x 80 or 6400lbs (say 3 tons) in the platten.

It was not therefore the heavy platten which was the offender but the high speed gearing. I set myself to design a drive in which none of the gearing except the rack pinion and the shaft on which it was mounted should be reversed. Without going into minute detail, I may say that I had two members mounted concentrically upon coil clutches on the rack pinion shaft. These travelled in opposite directions and had a ratio of five to one in their revolutions. The result was I was able to increase the speed in the cutting stroke from 12ft to 40ft a minute, and the cut from one eighth of an inch to half an inch or even more where necessary to bring a slab to an even surface at one cut. The usual process of planing was to take off all that was necessary to give the slab a flat surface, and then follow it with one stroke without feed as a smoothing cut, then turn the slab over, and do the same on the other side. I have seen a pile of slabs 4ft long, and the length of time taken to put them on the table, wedge them up, plane one face, turn them over, plane the other face and take them off was two minutes.

I admit that this was done as a show-off performance, but I have timed the operation over a considerable period, when the workman was not aware that he was being observed, and he was finishing a slab every three minutes.

A tremendous quantity of stuff that was being removed rendered special means necessary to get rid of it, and this was done by fitting a tilting extension to the platten, on to which the rubbish was delivered at the end of each cutting stroke, with the result that when the return stroke took place the rubbish was tipped over the end of the platten.

7. SOME ASPECTS OF THE QUARRYING INDUSTRY

About 1905 an exchange of territory with Rhosydd Slate Co had been arranged and finally agreed upon, and in this connection it was desirable that a very accurate survey should be made establishing the relative positions of the workings in the two quarries. I have always done my own surveying and I undertook to carry out this work. To make a closed survey and thereby check its accuracy it became necessary to run lines through about 1200ft of main tunnelling in the Croesor Quarry, about 300ft to the underlie of the vein thence obliquely back to the overlie, and across the strike of the vein to the 20th chamber from the main tunnel, up its working face and through a roof or drift, then through a very small manhole into the Rhosydd workings.

After this came five chambers, and up through Rhosydd open quarry for a distance of 400ft, then back over the main surface of the Rhosydd property, followed by the surface of the adjoining Croesor property over very rocky ground, eventually descending to the floor of the main tunnel where it joined the first line of the survey.

It was an extensive and very difficult traverse, but notwithstanding this the sum of all the angles checked right with not so much as one minute error. As regards measurements, the survey closed with only a difference of 0.6 of a foot in the northing and southing, and 0.3 of a foot in the easting and westing. The plotting was calculated to four places of decimals, lengths were taken by a steel tape, means being taken to prevent even the sagging of this. Parts such as the Rhosydd open quarry were measured by triangulation and means were taken to correct the very small error due to reflection of the light used for illumination of the crossed webs in the parts of the survey below ground.

I think that my surveying friends will appreciate the meticulous care necessary to obtain the degree of accuracy mentioned above.

I joined the Festiniog District Slate Quarries Proprietors Association as the representative of the Park and Croesor Quarries in 1896, and was elected almost at once as a member of the Executive Committee, in which capacity I continued to serve throughout the whole period of its existence. I was from the outset a strong advocate of propaganda in favour of slate, and among other things, the publication of a book containing information for the use of architects, slaters, and others. I myself as a geologist was the contributor of an article on the origin and formation of slate, most impressive figures being given as to its great geological age and the slow, but nevertheless crystallographic evolution of the material upon which its cleavability, elasticity, strength, and other good qualities depended. A great deal of time was devoted by the sub-committee to the preparation of this book, but I am sorry to say that it was never published owing to what I regard as a false idea of economy during a depressed period.

I was, however, able to secure the appointment of a lecturer to attend classes on building construction throughout London. One of the great difficulties in maintaining prices when trade was bad was to finance the holding of large stocks of slates when held by the less wealthy quarries.

Moreover, the inequality of stocks of particular sizes by different quarries was really surprising. One quarry could perhaps not sell its 14/7, and its neighbour had to refuse orders because it could not make enough of them. I brought forward a pooling system for all the quarries of the Association.

A pooling Company was to be formed in which the shareholders would consist entirely of members of the Association. The Company might have a large nominal, but a very small paid-up capital, and would take over into a common stock all the slates as they were made by the different quarries. When taken over they would lose their identity as the produce of any particular quarry, and the goodwill established would be the property of the Pooling Company.

Certificates in consecutive order would be issued by the Pooling Company to the quarries in the order in which the slates were delivered, and would be redeemed in cash as money came in from sales, in strict rotation according to their numbers. As in the course of trading, the time of redemption could be almost accurately predicted, these certificates would form a negotiable security, even although large stocks were held by the Pooling Company.

They would, moreover, be available almost immediately the slates were made. Quarry companies with ample funds would not need to cash their certificates before they matured in the ordinary way for payment for sales.

The pooling of stocks would very much facilitate the prompt execution of orders. A Committee sat to consider this scheme for months, but unfortunately when everything seemed ready to go ahead, one big company backed out, and the scheme had to be abandoned.

The Government were desirous of introducing more stringent rules for the use of electricity in mines. Their proposals were very far-reaching and very expensive to carry out. They required that every piece of electrical apparatus in a mine should be interconnected by an earthing system, and that this system in turn should be connected to an earthing system above ground.

Mr Nelson, the Chief Electrical Inspector, attended a meeting of the Association to hear criticisms and objections, if any, and I objected that the earthing at the surface of a mine was a redundancy, and involved unnecessary expense. The matter was referred by two of the companies to their consulting engineers, who disagreed with me, but nevertheless the Association appointed Mr RM Greaves and myself to go to London and put the matter before the Home Office.

We attended in due course and found ourselves confronted by a very formidable bench of experts, consisting of Mr Nelson, Sir Richard Redmayne, and a considerable number of others.

The presentation of our case devolved upon me, and I think I was able to prove that the earthing system at the surface would be inefficient, or at least negligibly operative, as no, or very little current would flow through it.

I instanced the installation at Croesor Quarry, where the transformer sub-station was below ground, and where a leakage from any particular phase would find its way back through the neutral of the transformer and thereby re-establish the balance of the phases. No current under these circumstances would reach the earthing system at the surface of the ground.

The bench of experts retired to consider my arguments and came back to ask me if the rule was deleted, would I consent to the substitution of a clause that the earthing system below ground should be open to the inspection of the Inspector, and this of course I accepted at once.

The North Wales Slate Mines were the only one of all the sections of the mining industry throughout the kingdom which secured such exception.

About the beginning of the Great War (1914-18) the Festiniog District Slate Quarries Proprietors' Association became merged in the North Wales Slate Quarry Proprietors' Association, which included the great Quarries of Caernarvonshire. The business of this Association was carried on almost exclusively by an Executive Committee of seven members, of whom I was elected one at the outset. I felt that the accountancy of slates was anomalous.

The prices were quoted on a nominal mille of 1200 slates, but was entered on the invoice as 1000. A tenth of this, or a nominal 100, was entered as 100.

In Caernarvonshire a quarter of a hundred was invoiced as 25, half a hundred as 50, three quarters of a hundred as 75, whereas in Merionethshire a quarter of a hundred was invoiced as 30, half a hundred as 60, and three-quarters of 100 as 90, but actually it was not 1200 slates that was despatched as 60 was added to the nominal mille of 1200, to cover breakage in transit, and a further 60 slates per 5000 was added also to cover breakage.

Actually, it worked out in practice that about 1272 slates per mille were being despatched.

I proposed to the Committee that we should cut out all these complications and substitute actual count for the nominal count hitherto employed. Also, making the units for slates 1000.

Any fraction of the 1000 to be entered as decimals, thus made 1500 slates a convenient entry in stock books or invoices of 1.5, eliminating tens of thousands of useless noughts from the stock books and other records. The Committee favoured my suggestions, and convened a meeting at Chester of the entire slate trade of the Kingdom, at which meeting I explained my plans, and suggested that the cause of the anomalous system was that slates were originally counted in dozens, and that instead of what is now called a mille of 1200, the count had been 100 dozens, which was logical.

The slate trade of England was represented by three Associations, the so-called National, which embraced the North of England merchants, the Midland, and the London Association.

The only real opposition came from the National, and it was discovered after the meeting was over that their opposition arose from a mistaken idea that the Welsh quarries wanted the trade to pay the same price for 1000 slates as they had hitherto paid for 1272.

The Committee decided that in the absence of unanimity the time was not ripe for carrying the scheme into effect, but after having twelve months in which to consider its advantages, the London Association took the initiative in asking the Quarries to adopt it. The Committee was, however, still reluctant to adopt it unless unanimity was secured, and communicated their decision to the London Association. The London body then informed the Committee that whether the Quarries adopted it or not, they would adopt it for the London area, and this was carried into effect.

At the first Annual Dinner of the London Association, I was invited as the original promoter of the scheme to become their principal guest, and had the honour of proposing the toast of the London Association.

Arising from the conditions of the lease from Lord Newborough to the Votty and Bowydd Slate Quarries Co Ltd, when Lord Newborough became a shareholder in the Company it had been arranged that the determination of any plant installed at the Quarry, being suitable and necessary for the carrying on of the Quarry, and also its value determined by and arbitrator, and failing that, agreement by an umpire, the parties concerned agreed to appoint me in both capacities as an umpire without reserving any right of appeal.

I carried out this work over a period of five weeks, and both sides expressed their complete satisfaction with my award.

I also for the Votty and Bowydd Co, made a complete inspection of the Yale Power Station, their transmission line, and all the electrical plant installed at the Votty and Bowydd Quarries, with a view to improving its efficiency.

I was consulted by the Dinorwic Quarries with regard to the electrification of the railway running from the Llanberis Quarry to the port of Dinorwic, but 3-phase current, though much to be preferred for most purposes, I found not to be very suitable for this railway, because of the complications rising at the two ends, and I advised the owners accordingly.

Acting upon the instruction of Mrs MA Owen, of Aberartro, I prepared an extensive hydro-electric scheme to be carried out on property, combined with transmission and distribution, over an area extending from Harlech to Barmouth. The granting of special orders by the Government had been suspended in view of pending legislation, but very great encouragement to proceed with the scheme was received from the Authorities.

A substantial amount of work on a dam was therefore carried out and pipe lines were laid.

Later on, Commissioners were appointed and they invited schemes to be put forward for the electrification of large areas by prospective contractors.

The scheme I prepared for the Barmouth-Harlech area was exceedingly well received by the Electricity Department at Gwydyr House, but was strongly opposed by the North Wales Power Co, who had put forward a scheme for the whole of North Wales, Chester, and Crewe, on the ground that the granting of orders for sub-areas would be inimical to the interests of the larger scheme. Although it was a great hardship and financial loss to the promoters of the Aberartro scheme, this view prevailed with the Commissioners and the necessary order was not granted.

I was also instructed by the Portmadoc Urban District Council to prepare a scheme for the transformation and distribution of electricity within their area, which included Portmadoc, Tremadoc and Borthygest. The scheme was approved by the Council technically and financially, but the Commissioners objected on the alleged ground that it was too expensive. A modified scheme was then prepared at lower cost and this was objected to also as being too expensive. A third scheme was then prepared, cutting down all cost to an absolute minimum, but still this was not accepted.

A distribution company, which was the offspring of the North Wales Power Co, put forward an alternative scheme, and a meeting, nominally for consultation with a joint electricity authority, was then called, but it became evident at this meeting that the Electricity Authorities were not desirous that sub-areas like the Urban District of Portmadoc should have separate orders, and as matters had apparently reached a dead-lock, the Portmadoc Urban Council withdrew their application, and the Distribution Company asked permission to substitute my scheme for the one they had themselves put forward.

About 1909 I went to Edinburgh with my wife on holiday and stayed at the North British Hotel. The morning after my arrival two men, strangers to me, came to the breakfast table and said: "Mr Kellow, we want you to give evidence regarding the Bay of Islands Quarry, Newfoundland." I replied: "I am here on holiday, have no desire to take on professional work, and know absolutely nothing about the Bay of Islands Quarry." "We shall subpoena you," they answered, "if you do not come voluntarily, but will you come up to our room in this hotel, and we can tell you all about it?" I replied to the effect that it was their evidence that I was expected to give, and not my own, but that I wouldn't mind accompanying them to hear what they had to say. On arrival in the room, I found a Scottish engineer going over his proof of evidence with counsel. It appeared that land had been leased to the Bay of Islands Company near the sea-shore on elevated ground, and that by inadvertence the intervening land between that and the sea, consisting of hillside, had been included in the lease, although the lessor of the Quarry property was not the owner of it. He, however, agreed to purchase it, hand it over to the Quarry Company, and thereby make good his contract, within a period of 12 months.

Unfortunately for him, he failed to do so, and the Quarry Company, taking the view that the possession of the hillside was essential to the success of the quarry, asked for the lessor to cancel the lease, and repay them the £6000 they had already expended on the Quarry.

This he refused to do, hence the legal proceedings. It has been one of the most extraordinary experiences I have had in my life, to find how prevalent is the idea that anyone, be he a grocer, a baker, or a candlestick maker, can learn all there is to be learnt about a slate quarry in a few hours. I have known men who had never previously been in a slate quarry, write a detailed report giving evidence of cost, output, and profit after having spent only a few hours in it.

In the case of the Scottish engineer referred to, his experience had been in Colliery work, and without doubting his competence as a colliery engineer, it appeared strange to me that he considered himself sufficiently expert to put forward a scheme for the working of the quarry by driving a tunnel from the bottom of the hill, and picking out the slate with hand picks. I could not help intervening, and told them that I knew the experts on the other side would completely upset this evidence. The upshot was that although I had not the slightest desire to do so, I was asked to take over the evidence as to how the quarry should be worked, which I agreed to do.

I was helped materially in forming my plans by a large model which had been erected in Court by the opposite side, and the Scottish Engineer, who had been to Newfoundland confirmed that it correctly represented the property.

When giving my evidence, I found that I had the Judge, Lord Guthrie, with me from the outset, and all attempts by the defendants' Counsel to break down my evidence failed. As I left the witness box, the leading Counsel of the Bay of Islands Company rushed up to me, shook me warmly by the hand, and said: "Mr Kellow, you have won our case for us! Although the claim first put forward by the Bay of Islands Company was only for £6000, the legal costs up to that time had already mounted up to £15,000, and were proceeding at the rate of £500 a day.

8. GEOLOGY

During my early youth I received very valuable training from my father in field geology, for which the mountainous district around Croesor afforded ample opportunity. Accessibility to the quarries also afforded facilities for examining sections of the strata and rock from which the surface layer of soil had been removed.

I did a lot of geological work at this time with Mr GJ Williams, who afterwards became the Assistant Inspector of Mines, and made the acquaintance of Mr Fearnsides, the very renowned geologist of Cambridge University, who proposed me as a Fellow of the Royal Geological Society of London, an honour I much valued.

I had made a collection of fossils, mainly from the Croesor district, although fossils were not very prevalent in the slaty strata. In this connection I was fortunate in discovering on the Garth Hill, near Minffordd, an uncontorted specimen of the Angelica Sedgwiski, which is a characteristic fossil of the period to which this hill belongs.

Every other known specimen in the world had been contorted by slaty cleavage, with the result that applications for casts of my fossil came from a large number of museums.

Eventually, as a sole return for the many kindnesses shown to me by Mr GJ Williams, I gave it to him to add to his collection.

At the same time, and as an allied science, I took up the study of petrology. Having a powerful petrological microscope fitted with analyser and polariser, and containing interchangeable eye-pieces, and objectives of different sizes. The maximum magnification used being 80 diameters, the most useful magnification, however, was much lower than this.

I used this microscope for the examination of rock sections cut thin enough to let light pass through them. These were mounted on glass slides, and covered with a circular piece of thin glass, the whole being fastened together with Canada Balsam which had the same angle of refraction as the glass.

I prepared slides not only of Festiniog, but also of Caernarvonshire and continental slate which I had cut in the plane of cleavage, in the plane of pillaring, and at right angles to both.

Slate, however, being of sedimentary origin did not present anything spectacular in the way of slides, but a number of additional specimens of igneous rocks with their pronounced and well developed crystals, which I also had in my collection, were not only interesting, but exquisitely beautiful, when examined under polarised light, and are a revelation to those who have not previously had the opportunity of seeing them.

My experience in geology and petrology served me in good stead when making examinations of properties in different parts of the world. Among these I may mention in addition to properties in Merionethshire and Caernarvonshire I examined and reported on properties in Cornwall, Devonshire, Westmorelandshire, several places in Scotland, including the great Ballachulish Quarries, and also on one property in Ireland.

Abroad I examined a French Quarry at Angers, where the method of working was to sink a very deep shaft and work the slate from underneath towards the surface, filling the cavity so formed with the rubbish.

I also visited a quarry in Luxemburg where the mechanical cutting out of slate rock was especially featured. I subsequently visited a very large quarry at Lichestein, in Saxony, where mechanical equipment of a very high order was in use. Eight hundred men were employed, and over two million tons of rock were being removed annually.

This quarry probably had established a world record by being made to pay its way with a yield of only 1% of manufactured slate to total rock quarried. (Whereas in Merionethshire and Caernarvonshire a yield of 6% to 7% is regarded as the economic minimum).

The three upper galleries of this quarry consisted entirely of waste rock. When I was there I saw a layer of 13ft in thickness being blasted away and removed. A line of holes was bored obliquely from the floor of one gallery to the floor of another several feet from the vertical side of the layer which was being removed.

These were fired simultaneously electrically, and a large mass of broken rock was rendered available for removal. This was then loaded by a mechanical shovel with a scoop that would take 5 tons at a time. A train of empty wagons, each with a capacity of at least 5 tons, was drawn by a steam locomotive along a track on the gallery floor. I never in my life saw work being done at such a rate. I stood upon the navvy while it was being worked, and timed a series of operations. The average time taken to fill a scoop with 5 tons of rubbish, swing it round, empty it into a wagon, and return for another charge, was only 35 seconds.

While the navvy was being swung backwards and forwards and loaded with rubbish, another empty truck was drawn up in readiness. I was informed by the proprietor that nothing was paid to the men for the first 1200 tons a day loaded, and that it was only after this quantity had been loaded that they began to earn any money.

There were other very interesting features in this quarry, including a slating school, where every variety of roofing was being taught. Pupils often working during the summer and coming back for further tuition in the winter.

I was in the USA for about two months, and during the time I was there I made a close examination of American methods of quarrying. To the casual visitor from this country the ramshackle buildings commonly found on the quarries gave the impression of a low standard of equipment, and it was only when one made a deeper study of the methods employed that it was possible to understand how the tremendous outputs per man were being obtained.

Briefly, the quarries are being worked on a pit system, and ropes are run across these pits, which are anchored at both ends. Along these supporting ropes carriages, fitted with a pair of pulley blocks, are traversed. Movable stops operated by the engine driver are pulled along the supporting ropes and stopped at any desired points. When the carriage encountered this stop, the wagon suspended from the lower pulley block would descend vertically. Except perhaps for the readily movable stops, there are many examples of this system to be found in this country, but the unique feature about those in American Quarries was that there were enough ropes, with their carriages and pulley blocks, each forming with its own engine a complete hoist, and so spaced as to make the distance between them small enough to enable a wagon to be lowered or raised from any square yard of the floor of the quarry. It was only the boxes of the wagons that were loaded into the pit, and never loaded above the sides of the boxes, could be loaded in a few minutes.

Even at American rates of wages, I found that the loading of rubbish at the bottom of the quarry, raising it to the surface, taking it to the tip and tipping it, was costing only 2½d per ton. No hand cranes were in use, as engine power was available for lifting big blocks, which in some cases I saw almost being dragged out of the rock itself. Blocks of 5 tons could be raised after one end had been lifted by a crowbar enough to get a chain around it.

At one quarry the splitters stood up when splitting their blocks, which were usually in three slate lengths. After being split down to the thickness of a single slate, they were tapped across with a chisel, and each pair of slates yielded six instead of two, as was usual in Wales.

The owner of this quarry was a relative of mine, and was staying at my house in Wales, when his quarry returns for the previous month reached him. He showed them to me, and I was dumbfounded to find that he was actually producing 16 tons of manufactured slate per man per month, against an output of about 2 tons per man per month at Blaenau Festiniog.

I am quite prepared for scepticism regarding this very high output per man, but from my own observation in the USA I can testify to the rapidity of production, and I can also testify as to what I saw with my own eyes on the quarry returns.

While in America I had the opportunity of visiting and inspecting a good many engineering works of considerable importance. At Niagara Falls I was conducted over the largest electric generating station that was there, by one of the engineers. It was one on the Canadian side, and their generators had vertical shafts, directly connected to the turbines underneath, and with field magnets revolving outside a stationary armature, forming very effective flywheels.

I was personally conducted over very important engineering works at Easton, Phillipsburg, Philadelphia, and other places, and in particular I visited the Niles-Cement-Pond Company's works there, the largest planing machine in the world. I am speaking from memory, but my recollection of the moving platten is that it was no less than 3ft thick, of solid cast iron, and a hundred men could stand on it.

There is no doubt that the USA is the land of big things. Another example of bigness is their stone-breaking machinery. Beside their monsters our stone-breaking machineries are pigmies. Manufacturing methods were converted into a fine art. Feeds, cuts, and cutting speeds were not left to the caprice of individual operators, but were scientifically determined by experts. Lathes were fitted with a large range of gears, and changes were simply made by moving a handle and did not require the antediluvian method then prevailing in England of taking off one lot of gear-wheels by hand and substituting another, the time occupied in doing which often made the change of feed not worth while. Spindle speeds determining the rapidity of the cutting speed were also variable by the simple movement of a handle. Capstan and turret lathes enabling a number of tools to be used simultaneously, and changes made from one tool to another, were in common use, and automatics, a large number of which would be under the control of a single man, were extensively employed.

Moreover, utility and not superfine appearance was the keynote of production. I saw cast iron pulleys and flywheels where the traverse of the cutting tool across the face was no less than ¾" per revolution, and the cut became a coarse spiral. No attempt was made to polish the work, and I was told: "We only turn it to balance it, and not to look at. If we are making machinery for Great Britain, we polish it, or we make it smooth and polish it, and charge a great deal extra for doing it." In a young country like the USA, imagination is not trammelled so much by precedent and habit, and therefore progress and efficiency have a wider scope.

While I was in New York I was entertained to dinner at the Engineers Club, and had the freedom of the Club given to me for a fortnight. Moreover, as the guest of the President of the Ingersoll-Rand Company, I had lunch at his Club, and my host pointed out to me in the same room many of the famous industrialists of the USA, including Mr Schwab, who was the General Manager of the Steel Company of America, and was in receipt of a salary of £240,000 a year.

9. ELECTRICITY PRODUCTION

Among the great hydro-electric power stations of the world that I have seen and examined, I may mention the one at Tyssedal, which is about 6 kilometres from Odda in the Hardanger Fiord in Norway, supplied with water from a natural lake 3½ miles long, and high up in the mountains. The lake had been dammed to act as a reservoir, and a tunnel through granite, nearly five miles in length had been driven, to act as a pipeline. At its lower end large diametered steel pipes of great thickness and reinforced on the outside by rings of steel of about 4" square section, had been placed at intervals of about a yard.

As the head of water was nearly 1400ft, even these pipes sometimes burst. I was shown over the power station by Dr Reich, who was in principal charge. A noteworthy unit in the station was a Pelton wheel, developing over 16,000hp from a single jet of water, and directly driving a 12,000kva alternator. This set was of German make, and was supplied during the Great War in consideration of the British Government consenting to Germany getting nickel from the other side of the fiord.

Dr Reich said that the Germans made the condition that the set was not to be used for the work of the Allies, and he replied: "Gentlemen, I will do the best I can, but of course if any of the current should by accident leak into my bus-bars I cannot help it!" I thought that the suggestion the current from a 12,000kva alternator should accidentally "leak" into the bus-bars was very comical.

"Of course," said Dr Reich, "as soon as we got the set delivered we put it to work for the Allies." The power developed at this station was 180,000hp.

Above the reservoir-lake supplying this station there were a number of waterfalls, one of which was the grand and majestic Schededelfoss, 525ft high! On the day following my visit to Tyssedal, I went to see the famous Lotefoss and Skarsfoss Waterfalls, which derive their water from a single river, fed from Lake Reinsos, and supplemented by water from the Krossfonn Glacier.

The river divides just above the two falls, and joins again immediately below them. These falls are regarded as among the most beautiful in Norway, but in my opinion are outclassed by the majestic Schededelfoss, to which I have already referred.

To return to the Tyssedal Power Station, the current is used for the manufacture of carbide of calcium and other industrial purposes. It also supplies the town of Odda with electricity. I found here a tariff in force based on a different system to anything I had been previously acquainted with.

The right to use current for 24 hours a day for 365 days in the years was given for the equivalent of about £6 sterling, which worked out to about one-sixth of a penny per hour. Of course, it is unlikely that any consumer would be in the position of using it without interruption, unless it was for power purposes, but even if it was only used half the time, the cost was only one-third of a penny per unit for all purposes—lighting, heating, cooking, and power—which of course made it cheap enough to use in substitution for any kind of fuel.

I asked what happened if a consumer turned on more current than he was entitled to, and was told that this was nullified by the use of current limiters which reduced the voltage so as to keep the wattage within the agreed limit.

The other hydro-electric station which I inspected was at Trollhatten, about 7 or 8 miles from the Swedish town of Gottenburg, linked up with the Gottenburg Canal, along which I sailed, which was nothing short of an inland sea requiring more than a day to cross it by steamer. This inland sea formed the reservoir for the power station at Trollhatten.

A large number of mixed flow turbines directly driving alternators were fixed in line and could be seen from a gallery at the end of the room containing them. The head of water was not great, but the volume was immense, and the water issuing from all the turbines at once when they were at work, was sight to behold. A veritable ocean of water! There is little doubt that Norway and Sweden possess great resources in water power, in comparison with which most of the water powers in the British Isles are small, but the North Wales Power Co's installation at Cwm Dyli and Dolgarrog, in Caernarvonshire, and the Scottish hydro-electric plant at Kinlochleven are of quite respectable proportions.

I visited on the continent of Europe the works of electrical manufacturers at Charleroi, Helois Works at Cologne, Allegemaine Electricitats Gesselschaft's works at Berlin, where they employed 18,000 men and ran the whole works without overhead shafting or belting, every individual machine being motor-driven.

This left the whole of the space overhead available for overhead travelling cranes. All the bays could be seen from an elevated platform near the centre of the works. The Offices were so extensive that on the first floor there was actually a tram car running in one of the corridors to convey passengers from one office to another or to the observation platform in the centre of the works.

I also went to the Schukert Works at Nuremburg, Kolben's Works at Prague, Oerlikon Works at Zurich, Brown Boveri's Works at Bale, and the works of Messrs Sautter Harle et Cie of Paris.

I also visited important works other than electrical ... Sulzer Bros at Winterhur, and the turbine works of Escherwyss, who were the makers of the Niagara turbines. At about the same time the great Westinghouse works at Trafford Park, Manchester, were in course of erection and nearing completion, and I was conducted over these by the General Manager of the Company.

Later on, in these works, I saw the Parsons steam turbine being manufactured. This was of special interest to me, as at the very inception of steam turbines I had the honour and privilege of being personally conducted over his works at Newcastle by the inventor, Mr Charles A Parsons (afterwards the

Hon CA Parsons). He was then conducting his famous experiments with the Turbinia, and he explained to me that at first they could not get the speed of the boat up to what was expected, and found that this was due to cavitation. To explain this more fully, I may say that the turbine, which was directly connected to the propeller, turned the latter so fast that it displaced all the water in the space in which it revolved, and there was nothing for the blades of the propeller to push against.

A reduction in the speed of rotation rectified this and made the Turbinia a success. When I saw the tiny machines mounted on a bench and realised that they were capable of developing 1000hp, I realised the vast potentialities of the invention when full-scale production was being obtained.

Later, at the British Thompson-Houston Co's works at Rugby I saw a Curtis turbine, which is a modified form of the steam turbine involving large diametered impellers, but with fewer stages. In any event, I regard the steam turbine, with its compactness, high efficiency, and the enormous power it is capable of developing as being one of the most revolutionary inventions ever introduced for power production.

Very shortly after my visit to the works of Mr CA Parsons at Newcastle, I installed at the Croesor Quarry a Laval steam turbine running at 24,000 revolutions a minute, and developing 10hp The speed was reduced within the machine itself by double helical, very finely cut gears to 2400 rpm at which speed it was directly coupled to a ventilating fan.

The Laval turbine worked with only a single jet, and a single stage, and hence the very high speed of rotation rendered necessary. My reason for using this turbine was because of the small powers in which it was procurable. It did its work quite satisfactorily.

10. MY SON

I decided to bring up my son as a Mining Engineer. He was educated at the Advanced School at Portmadoc, under Mr Rees Evans, up to 13¾ years of age. I then sent him to Dulwich College, London, as a preparatory College for Cambridge. He here had the distinction of being House prefect and School prefect, and in sport was Captain of the Cricket XI. He subsequently entered Cambridge and took the Mechanical Science (Honours) Course. He was a Member of the Cambridge Engineering Society, which was common to the whole University.

Papers were read at the meetings of this Society by outside experts in all branches of Engineering, but during the year the President of the Society selected three undergraduates, whom he deemed worthy and competent to read papers before the Society. My son had the distinction of being selected as one of these three; not only that, but he was awarded the Prize for the best paper of the year. It consisted of a surveying aneroid, enclosed in a gold case, on the back of which was inscribed:- "CUES Prize, 1913. 'The Drilling of Rock' W Kellow, Caius College." The gold case was in turn enclosed in a leather case.

From Cambridge my son went to South Africa, where, after a very short time, he was appointed Surveyor of the Ginsburg Gold Mine, near Johannesburg. During the Great War he applied to the Company for leave to enlist, but was persistently refused on the ground that he was an essential man. Where leave was granted to enlist, the Company continued to pay the salary, but if anyone enlisted in defiance of refusal by the Company, no salary was paid.

Notwithstanding this forfeiture of salary, he enlisted in the Royal Flying Corps in South Africa, came to England for training, and subsequently went to the war in France. He was shot down and killed on the 16th of September, 1918, shortly before the Armistice was signed. As he was our only son, and only child, my wife never recovered from the shock, and died literally of a broken heart in August, 1919.

I think that the time that followed this represents the lowest ebb in my life, as I had looked forward to the time when after an experience in important mining undertakings abroad, my son would be able to return to this country and assist me in my activities, and eventually take my place.

Within a few months after I had lost my wife, I was approached by a group of four companies, all of world-wide reputation, who had entered into a combination to promote hydro-electric schemes throughout the world, and they offered me the position of being their Consulting Engineer.

I was 57 years of age at the time, and with the loss of my wife and son, the incentive to undertake work of such responsibility and magnitude was lacking, so I refused. My refusal was only accepted as final after I had informed them four times that I could not see my way to accepting their offer.

I consented, however, to examine one scheme for them, and one of the features of this was a dam which would cost £100,000 to construct, but on my informing them of this amount the figure did not in the least stagger them. I confess that it was a surprise to me that I, living at Croesor, should have been selected by so influential a group when there were so many eminent hydro-electric engineers at Westminster, from which a choice could have been made.

11. MY CAR AND ITS PETROL CONSUMPTION

In March, 1920, I purchased a Morris Oxford 11.9hp car, which the makers estimated would do about 23 miles to the gallon. I think I am almost incapable of accepting and using anything without a desire to improve it, and this car was no exception.

My efforts were directed chiefly in improving the petrol consumption. The usual method of attacking the problem was to reduce the size of the jets, but this was accompanied by loss of power, overheating of the engine and radiator, and more difficult starting. Therefore, the project was abandoned as not being worth while. After many abortive attempts, continued over some years, I was successful in achieving what I set out to do in a remarkable degree.

One of the first conditions of success was to eliminate the haphazard methods of measurement, and I therefore purchased a Hamil Flow-meter which was a highly scientific instrument, showing accurately at a glance the exact rate of flow of petrol per hour. I mounted this on the dash-board side by side with the speedometer, so as to be able to observe the two instruments simultaneously, and by a rapid mental calculation ascertained the number of miles per gallon the car was doing under the conditions prevailing at the time.

The effects of speed variations, gradients, etc., could be instantly noted, and did not depend on measurements of petrol taken under difficult conditions at the beginnings and ends of journeys. Notwithstanding the difficulty of the latter, I did however, check the momentary result given by the flow-meter and speedometer readings, with a total consumption over long journeys. I may say that so far as petrol consumption was concerned, I approximately doubled the mileage per gallon.

On a long tour starting from Croesor in Merionethshire, going and returning through the Lake District, and taking in almost the whole length and breadth of Scotland, I obtained some remarkable results. I travelled from Perth to Aberdeen, stated as 70 miles, on an exact measured gallon of petrol. I later travelled from Inverness to Perth, 114½ miles, on two gallons of petrol, and during the whole tour the average worked out to about 50 miles to the gallon. I could travel along the Llandudno sea-front, on a good level road, at 20 miles per hour with a consumption of only 2 pints per hour of petrol, which works out to 80 miles per gallon.

So far from there being any loss of power, there was a definite increase, and notwithstanding the jets being reduced to motor-cycle sizes, the car had been converted into what was practically a top-gear car. With a clean or decarbonised engine I could travel from Croesor to my hotel at Llandudno taking the Aberglaslyn hill, the hill from Nantgwynant to Pen-y-Gwrhyd and Bodnant, and all the other steep hills without changing gear.

I kept a perfectly cool engine and no unsatisfactory symptoms of any kind manifested themselves. Moreover, an engine that had been difficult to start, would start at the first touch of the switch, without even throttling the air in the carburettor or closing the air valve. I am fully aware of the extraordinary nature of the claims put forward, and I anticipate that they will be received by others with the same scepticism as I myself would have displayed did I not know the facts.

As, however, I knew that to exploit this was not a side issue but a full time job involving all sorts of risk and expenditure that I was not prepared to undertake, I decided to use it privately, and to defer until some future time the question of utilising it in a more extended sphere.

In 1940, and from patriotic motives, I decided to inform the Ministry of Supply of the results that I had obtained, but I made it perfectly clear at the outset that in my then low state of health I was not prepared to undertake the laborious task of patenting it, nor to incur the heavy expense of patent counsel without which no satisfactory specification would be obtainable.

Moreover, I emphasised the point that I was not prepared to advance money which might or might not be repaid at some future and probably remote time. Knowing something of the routine of Government Departments, I was not surprised to receive a form marked let it be said "ABC/XYZ No.40," in which the only conditions acceptable to the Ministry were that I was to prepare two sets of drawings, with full explanations of the principles and methods of working my invention, at my own risk and expense, and if found satisfactory by the Ministry of Supply, they would still pay no money until it had proved itself in actual practice in the Government services to be of tangible value.

A relative of mine, who had rendered great service by his inventions during the Great War, had not been paid anything until after the war was over, although he informed me that there was an accumulated liability amounting to £90,000. One or two more letters to the Ministry, and their replies thereto, convinced me of the inflexibility of their rules, and so I dropped the whole matter.

I do not blame the Officials, for I recognise that in matters carried out on a national scale rules of procedure are necessary which may often operate in a manner detrimental to the Country's interests, and this inflexibility is in my opinion one of the great objections to the running of the Country on a collectivist, as opposed to an individualistic system, which latter gives more latitude for modifying a course of action, when such modification appears to be advantageous.

In March 1920 Moses Kellow purchased a Morris Oxford 11.9hp car

12. FARMING AND COLLIERY INTERESTS

The Croesor Quarry, when started in 1895 was a leasehold of 240 acres. The royalties had been purchased for a long period of years, after which onerous terms came into force.

Unfortunately, working at the Quarry had been suspended for 17 years of the royalty-free period, leaving only 10 years of the period unexpired, an insufficient time in conjunction with the onerous terms thereafter applicable in which to repay shareholders for the capital outlay, required for its development.

Shortly afterwards, I was successful in arranging terms for the purchase of the whole Croesor Estate of 1343 acres, together with the Croesor Railway on both the Croesor and Park Estates. Some time later, I also arranged the purchase of the freehold of the Park Estate, the combined area of the two estates amounting to 2700 acres. The management of these estates and of the Croesor Railway were in my hands up to the final sales of the properties.

The Portmadoc, Beddgelert & South Snowdon Railway which linked the Croesor Railway with Portmadoc, was only under my Management during the Great War.

As a further agricultural interest, I may mention that I managed a farm of 160 acres in Caernarvonshire for my father, who had gone to Cornwall as the General Manager of the Delabole Quarry, for a period of three years. I thus had experience of estate management, as also that of running a farm in its various seasonal activities and ramifications.

About 1898 I visited South Wales and with a companion went down the Windsor Pit at Ynysbwl, which was the property of the Ocean Company. I took the opportunity of going down when the night shift men were being lowered to their work about 6pm.

The pit was not only one of the deepest in the country, but had a winding gear which was of the most modern type, and very rapid in action. The descent gave one the feeling that the earth had literally dropped away from beneath one's feet, and when deceleration took place near the bottom the sensation of pressure was intense, no trip for one suffering from cardiac trouble.

At the bottom we were taken charge of by a fireman who was going on his rounds of the Colliery. The roadways were flanked on both sides by lines of pit props which had been splintered to matchwood by the approach of the floor and roof owing to the super-incumbent strata, at least half a mile thick. It was disconcerting to the uninitiated but it did not worry me much owing to my long experience in underground work.

I was surprised at the skill that was manifested in the lay-out of the roads—up inclines, down declines, round sharp corners—and how long trains of wagons could be pulled along such tortuous tracks without derailment was a marvel.

At one point in our journey, our conductor said to us: "I do not know whether you will be able to follow where I am going now, but I warn you it is not an easy matter." Only the fireman had a lamp, so we had willy-nilly at least to attempt to follow him or stay there in total darkness by ourselves. We decided to try and follow him.

Without further word he pitched himself head foremost into what appeared little more than a good-sized rat-hole, inclining about 45 degrees. I followed! The hole was so close a fit that I could not bend my knees, and it was only by a wriggle, aided by the slope, that any progress could be made. When I was about half way down, there was so little room to spare that I seriously feared a jam, and my position, with legs upwards and head downwards, with half a mile of earth above me, and in total darkness, was definitely an uncomfortable one.

The tour of the mine took so long that we did not return to the pit bottom until about midnight, where we found about fifty men sitting down. Enquiring what was the matter, we were told that about two hours previously something had gone wrong with the winding gear, but what it was no-one knew, and there was no telephonic communication between the surface and the pit-bottom at that time.

We asked when winding was likely to be resumed, and we were informed that it might be very soon, or it might even be several days, and that nothing could be done about it except to sit down and wait. Winding, however, re-started at 2am, and we were able to return to two very worried wives, whom we had not advised of our intention to visit a Colliery.

I next visited the Powell Dyffryn Colliery, where an experimental plant for the extraction of petrol from coal had just been installed. I think it was the first that had ever been introduced into Great Britain. It had been made, and was I believe operated by a German firm.

Incidentally, during this visit to South Wales, I met Mr DA Thomas, the Colliery Proprietor, who said to me: "I think that you live very near Mr Lloyd George, do you not?" "Within eleven or twelve miles," I replied.

"He is a very smart young fellow," said Mr Thomas, "and mark my words, he has a great future before him." This remark was made long before Mr Lloyd George had attained fame, and in the whirligig of time how true the forecast proved was indicated by Mr Lloyd George becoming Prime Minister, and Mr DA Thomas, later Lord Rhondda, serving under him as Food Controller.

While on a visit to Rose Warne, Camborne, Cornwall, the beautiful home of Mr James Holman, the Chairman of Directors of Holman Bros, whose guest I was, my host took me down the Dolcoath Tin Mine, which is one of the largest and deepest in the world. I was told that more than 2000 years previously the Romans, at the zenith of their power and industrial expansion, had worked this mine for tin, and that its profits during this long period had been fabulous. With increased depth and enhanced cost of working, the tin industry of Cornwall had become unprofitable, but my host was of the opinion, even then, that with the increased scarcity of tin throughout the world, and the increased prices then obtainable, there was a good prospect of the industry being re-established upon an economic basis.

About 1900 a company called Kellow & Co Ltd, of which I was the Chairman, was formed to carry on a wholesale and retail coal business at Portmadoc, taking over a private business of the same name at Portmadoc. At that time extensive shipments of slates were being made from Portmadoc to South Wales, and the slate vessels returning from South Wales were glad to have coal instead of ballast for their return journey to Portmadoc.

Cheap rates for these coal cargoes were obtainable, and an economic basis was thereby rendered available for bringing steam coal to Portmadoc for distribution to the various slate quarries at Blaenau Festiniog.

The electrification of the quarries, by dispensing with the necessity for steam coal, spoiled this market, but foreseeing this, a wholesale rail trade had been developed in the adjoining counties, in addition to a very substantial retail trade. Nevertheless, my other activities were making such a demand on my time that I thought it desirable to sell my shares in the company.

13. MY TRAVELS IN AMERICA

Up to this point I have confined myself to a description of my activities in technical, professional, and business directions. Now I will deal with some of my experiences in other directions.

I have travelled extensively in the British Isles, and there are few towns or places of interest, beauty spots, or historic sites which I have not visited at one time or another. I have seen nearly all the great Cathedrals, castles, museums, picture galleries, universities and public schools, and have seen in the latter the names, cut in the desks, of a large number of boys who afterwards rose to great eminence in the world. I was particularly fond of the Tate Gallery and the South Kensington Museum, but I will not occupy space by going into detail concerning things and places that are common knowledge to most people in this country.

In Ireland I made one tour from Dublin through Killarney to Cahirciveen and the Island of Valentia. After my return from a very extensive tour, passing through the Lake District, and the throughout Scotland, I composed some verses (I will not claim the title of "poetry" for these!) commemorating the trip. The first relates to the Island of Skye, and runs as follows:

When we were little kids at school,
  And taught about the sky,
We thought 'twas made up of thin air
  In which the birds could fly:
But here we found the Skye so dense
  The birds could walk upon it,
And Highlanders upon it dwelt,
  Sporting their kilt and bonnet!

The second example relates to Perth, where the great Dye Works are situated:

And then at last we came to Perth,
  Of paradoxes trying,
Where the pushful folk were Pullars,
  And the people lived by dyeing!

The third and last example I shall quote referred to the Morris Oxford car, which I have already alluded to:

And now about the little car,
  Of Horse Power 11.9.
It did its 50 miles per gall,
  A record, I opine!

On my journey to America, the liner Kaiser Wilhelm der Grosse, which had taken from us the Blue Riband of the Atlantic, passed us very close on the larboard side, and the German passengers gathered on her decks and cheered contemptuously as we were in earshot. Just before entering New York Harbour we again saw the Kaiser Wilhelm der Grosse—stuck fast on a mudbank—and it was we who passed her this time. The cheering was returned by us, with interest.

I might here mention that during my voyage across the Atlantic, the Chief Engineer very courteously offered to show me over the ship, and gave me an opportunity of inspecting its mechanical equipment, which offer I gladly accepted, as marine engineering had not previously come within the ambit of my experience. I think I was the only one of over a thousand passengers to whom this honour had been extended. I did not know the Chief Engineer personally, but I think he had heard of me through my institutional connections.

I found New York an impressive spectacle as I entered its harbour for the first time. Its marvellous collection of tall buildings silhouetted against the sky, were the New Yorkers' solution to the problem of finding space. But the city, I soon discovered, was striking in another way, laid out as it was on a chess-board plan, with avenues running one way and streets crossing them at right angles. Both avenues and streets were numbered consecutively—The Avenues from Broadway, and the Streets from the Water Front.

Every block consisted of 100 houses so that, for example, house number 1061 would mean the 61st house in the 11th block, and 1601 the 1st house in the 17th block. This applied both to avenues and streets, and if an address was known it could be easily found or its position mentally visualised. With elevated railways running the whole length of the city in every third avenue and street cars on the ground level, travel to any point was very simple and convenient.

Apart from the skyscrapers I did not think the streets of New York were good to look upon.

Very tall houses alternated with low houses, and as they nearly all had flat tops, they reminded me of a lot of packing cases placed side by side.

I passed several times over the bridge between New York and Brooklyn, where it is claimed that the traffic is the densest to be found on any bridge in the world.

A description of New York would not be complete, unless its playground was also mentioned. This was Coney Island, where sea bathing and other aquatic sport could be indulged in, and where was located, I think, the maddest and most extraordinary collection of amusements that man had devised.

It cost about £3 to view all the side-shows, and they ranged from baby incubators to a giant mangle, between whose rollers a wide belt passed endlessly. Some distance back from the rollers men, women and children were being dropped down on this moving belt, and were carried towards the rollers to what seemed certain death. Of course, they emerged on the other side quite unhurt, although their immunity was not obvious from the entering side.

One of the side-shows had a Negro seated on a stool, with a thick cushion round his neck, at a barrier only a few feet distant from the spectators, whom he was challenging in a bantering tone to hit him on the head with wooden balls.

The visitors as it happened contained several cricket players, who could throw a ball not only straight, but at lightning speed, yet with apparently only the slightest movement of his shoulders, the Negro was able every time to dodge the missile so that it hit the cushion quite harmlessly, while he constantly chaffed the throwers by remarking how much he enjoyed his unusual experience, and that he would not change seats with the President himself! An amusing incident occurred in connection with the Baby Incubator, which might bear repeating. I had taken a man with me to America, and I had impressed him somewhat with the vastness of the country, and how essential it was to the welfare of the United States that the population should reach a certain density, so that roads, railways, etc., could be provided and worked economically for the benefit of all. I told him that the Americans were a very inventive and mechanically minded people, and as the required density of population was not being obtained, they were now producing population by machinery.

He very naturally and correctly thought I was pulling his leg, so I showed him the Baby Incubator and we both entered its inviting portals with due solemnity.

"Now see for yourself," I said. "Now will you believe what I told you?" After a long pause the cautious answer was: "I have seen the babies, but what is puzzling me is, where they came from in the first place!" During my stay in New York I learned several things: not to stop a policeman in the street to ask him questions; not to hire a taxi in New York unless you are a millionaire or nearly one; not to put your boots out to clean at the bedroom door of your hotel at night (or you may not see them again!); not to forget that the hotel servants are much your superior; not to forget that a dollar is a common tip for quite trivial services and that a quarter is the minimum ever offered; not to forget that you can buy nothing for less than a nickel ($2^1/_2$d), and that cents are only used to put in the collection plate on Sundays. I write, somewhat facetiously, of the time when I was in the USA.

At Philadelphia, besides other places of interest, I visited the renowned Wanamakers Stores, which I believed was then reckoned to be the largest in the world, and one of Philadelphia's showplaces.

At Washington I visited the White House and at a short distance farther up the hill the splendid Congressional Library, the entrance hall of which, including the stairs and gallery, is all executed in beautiful marble and is, I believe, a replica of the entrance hall to the Opera House in Paris.

The houses in Washington—or as the Americans call them, the "homes", which I consider is a pleasanter description—are much nicer than those of New York, space admitting of more ascetic design. During one of my trips through the city in a sight-seeing automobile, we passed the statue of General Washington, and this was what the Conductor said: "This is a statue of General Washington. The statue cost 20,000 dollars and the pedestal 15,000 dollars—a total of 35,000 dollars. Move on!" He did not add a word of description about the statue, not about General Washington, or anything else, but confined himself to a summary of the cost. I looked round at the other passengers on the automobile, and they appeared to be completely satisfied with what the Conductor had told them. I mention this, as I found it to be typical of the American mind. The point of principal interest appeared to be the dollar value of everything, including human beings. The aristocratic are not Dukes, or Earls, or Viscounts, but those who are possessed of most money, and I have been astonished on meeting complete strangers at being informed by themselves, within a few minutes of being introduced in the most informal manner, that they were 500,000 dollar men, or 100,000 dollar men as the case might be. I have had the same kind of information communicated to me by letter from people who have been writing to me for the first time.

As I mentioned, even human life is assessed in terms of the dollar. I took an extensive motor tour through the States, and soon found that the highways often ran side by side with the main railway tracks, often also, zigzagging across the tracks according to the limitations of space, and that with no fences or gates by way of protection.

I remarked on the danger of this to my companion, and he frankly admitted: "Oh, yes—we have 80,000 accidents a year on our railways, against less than 10,000 on yours, but the railway directors have gone into the matter very carefully, and have found that it is cheaper to have the accidents and to pay compensation for them than it is to put up fences." He said all this without apparently having any doubts as to the wisdom and propriety of the directors in making their decision, but let us not be too censorious. I can remember the time, not so long ago, when our bridges below ground had no hand rails, and dangerous shafts were left unfenced in complete darkness—regular death traps, because we did not want to spend money on fencing.

I went about a good deal in the States in the company of Mr William Davies, the brother of Mr Richard Davies, and Mr Jonathan Davies of Portmadoc, who showed me very great kindness.

One day, when I was at his home in Danielshill, we walked out into his garden, and I noticed steps descending into the ground, and asked him where they led to.

"Down to our cyclone cellar," he replied. "If we look around, and see trees, houses, and all kinds of other things being picked up and carried through the air, and their progress happens to be in our direction, we know that we have a cyclone coming, so we leave the house and retire into the cyclone cellar." The ground is made perfectly level above the cyclone cellar so there is nothing for the wind to take hold of. In the calm peaceful summer afternoon it seemed difficult to think that such a calamity could occur, bringing death and destruction to a wide district, and it made me feel that after all we had in Britain some things to be thankful for.

While in New York I visited the Stock Exchange in Wall Street. The public are admitted to a gallery overlooking the floor of the house where the members conduct their business. This gallery has a railing in front, surmounting a small curb about a foot in height. When I entered the building I was carrying a tube containing plans, but the attendant stopped me, saying: "You can't take that in there!" "Why not?" I demanded. "It contains nothing but plans." I opened the tube to show him its contents, but he was still somewhat reluctant to let it pass.

"Well," he said at last, "I'll let you take it in, provided you keep it quite level, and hold it below the line of the curb in the gallery, for if the members caught a peep of it through the railings there would be an immediate stampede as they would mistake it for a bomb." So many people had lost their money there that there were always a few, unhinged by their losses, who sometimes attempted to revenge themselves by blowing up the place. Needless to say, I was careful to carry out the attendant's instructions.

The scene below on the floor of the house was one of feverish activity. Everyone seemed to be almost running about, and there were tens of thousands of tiny shreds of paper littering the floor, the remnants of torn-up notes—torn into such minute pieces that no-one could read anything on any one of them, and thereby gain any information that might be useful to him.

Every nation has its characteristics, and the Americans have many that are admirable. They are very hard workers, and not time wasters, are inventive, very progressive and hospitable to a degree. I shall never forget the kindness I received at their hands.

In New York I stayed with my cousin, Martin Kellow, at his beautiful home in Warburton Avenue, Yonkers, which is a large residential area near New York. His father, Joseph Kellow, who was then a very old man, told me that he was the first one to open Slate Quarries in Pennsylvania, and that this was done in what I may call the Bangor-Penargil and Windgap area. The various members of the Masters and the Kitto families have also figured largely for generations as owners and operators of Quarries in this district, and within the last eighteen months I learned with regret of the death of my cousin, William Kitto.

Prior to his death, for a long period of years he had been General Manager of a combine of quarries (21 in all) producing structural slate.

The trade catalogues running into several large volumes contained numerous beautifully coloured illustrations, model specifications for almost every class of slate work, and other useful information. The catalogue exceeded by ten-fold anything I have seen elsewhere, in excellence and comprehension, having to do with the slate trade.

I think therefore that my relatives—the Kellows, the Masters, and the Kittos, have for generations taken a notable part in the American slate industry.

During this American visit I made the home of my cousin, Thomas Masters, in Penargill, my headquarters, and travelled by a famous train which, if I remember rightly, was called the Diamond Express, to Niagara Falls. It is common knowledge that these falls are in two parts, the American Falls in which the cascade is a straight line, and the Canadian Falls, which are in the form of a horse-shoe.

They are divided by an island known as Goat Island, and it was from this island that Blondin wheeled his assistant in a wheel-barrow on a rope suspended immediately over the rapids above the falls, where, of course, a slip would have meant certain death.

Some years later I saw Blondin wheel his assistant along a rope, and several other even more daring exploits at the Agricultural Hall in London! About thirty yards on the American side of Goat Island a rock juts out at the top of the falls which makes a little gap in the torrent of water pouring over the edge. It is the space between this thirty yards wide fall and the rock wall behind, which is known as the Cave of the Winds. I visited this Cave.

I was divested of all my clothing on Goat Island and clad in oilskins and sou'-wester, the latter tied firmly to my head. I was then lowered on to some rocks, which were almost level with the seething pool at the base of the fall. Travelling along these rocks, and along a very crazily constructed short length of planking, I was able to pass through the gap between the narrow and main American falls. It was necessary, however, to go so near to the falls themselves that huge splashes of water falling from a height of 180 feet came down on one's head. I was nearly knocked over by one of these, and the wind was so terrific that though my sou'-wester was tied down to my head with a rope, it was blown to the back of my neck, and in the instinctive effort to catch it, I raised my hand without realising that my sleeve was full of cold water, and this, of course, was poured all over my body inside the oilskins.

The journey was continued behind the falling water in a hurricane of wind, until we emerged at the other end, and thus completed the circuit. It is not an experience I recommend to a nervous person or to a lady.

On the American side of the falls there was a little steamer known as the "Maid of the Mist", which sailed as nearly as it dared to the base of the falls. The seething water in the pool was so full of air bubbles that any careless boat or steamer might easily sink in it, but I presume those in charge knew where they could venture to go, and where not.

In any event, the tossing about of the boat on the turbulent water was an experience which I remember vividly to this day. From this pool at the base of the falls, the Niagara River rushes down through a gorge six miles long to a whirlpool. I went down one side of this gorge, and came up the other. When I came to the whirlpool I did not notice much evidence of whirling, but while I was looking at it a baulk of timber about 16ft long came floating down the river towards the centre of the pool, and to my intense surprise, suddenly assumed a vertical position, about half its length being up in the air. Then, abruptly, it shot out of sight, sucked down by the vortex action in the centre of the pool.

Shortly afterwards, I returned to England from New York, accompanied by my cousin, Mrs Thomas Masters, her cousin Emily Kitto, and Mrs Masters' baby, Walter, four months old. It is typical of the quickness of decision which characterises Americans that it was nine o'clock at night, preceding the day on which I intended to travel from Penargill to New York, that the idea of accompanying me was first mentioned.

I made the remark that I would have liked my cousin, Mrs Masters, to accompany me, without thinking that it was then a practical possibility, and she answered: "I would go, if I could persuade Emily to come with me, to help with baby." So Emily, my young cousin of eighteen, was telephoned, and within an hour of the proposal being first mooted, had decided to come to England.

Bear in mind, no passages had been booked, no special clothes bought for the journey, or any other preparations made, yet next day we started blithely for New York. When we saw the booking clerk every berth had been booked, but we were told to call again at 11am, on the chance of there being withdrawals. We were lucky, and at mid-day we were on the boat, bound for England. This gave me the opportunity of repaying in kind in some degree the hospitality I had received from Mrs Masters and her husband while I was in America.

I took them to London and elsewhere and showed them sights that were characteristically British, but I shall never forget a remark they made to me: "What we want to see most is Kings and Queens—that we have not got in America!" I have found this to be a common desire among Americans visiting this country. Later on, as he grew to manhood, the baby Walter achieved national fame as a baseball player.

14. AGRICULTURE

In 1932 I paid a most interesting visit to the Channel Islands, travelling to them from Weymouth, and returning via Southampton. The boat stopped at Guernsey, but I did not disembark until we finally reached St Helier, the capital of Jersey.

This Island may be called an island of glasshouses supplying a stupendous quantity of flowers and hot-house fruit for the English market. Jersey, the principal island of the group, I found to be one of the most interesting places I have ever visited. The coastal scenery is superb. It is almost surrounded by rocky islets upon which the waves break in foaming whiteness, and when the sun is shining brightly down upon the sea these islets are like jewels in a beautiful setting.

Among other things characteristic of Jersey are the famous Jersey cabbages, which grow to a height of several feet and are used as cattle food. The stems are often made into walking sticks.

Visitors are pretty sure to be asked a variation of that time-old riddle: "Why is a Jersey cabbage like a mouse?" to which the somewhat obvious answer is: "Because the cat'll eat it!" The island is of course also noted for its wonderful breed of cows, whose milk is 2/5ths pure cream. These cows are so tame that if you stand anywhere near them they will walk up to you and push you with their noses.

Near the northern shore of the Island is a camera obscura, where the man standing at the door invites you to come in and see the whole of Jersey and England! All for 3d.! You enter, and in the camera obscura you see the whole of Jersey, but not so much as a thin shore-line of England. You remark that you had been promised a sight of England as well as Jersey, whereupon the wily man in charge says quite proudly: "Well, have a good look at me. I am Mr England!" However, quite the most important thing about the Island is its agriculture. The Island is only 8 miles long by 5 wide, and yet when I was there, there were no fewer than nine steamers loading potatoes. I observed a queue half a mile long of wagons and carts loaded with potatoes, all waiting their turn to pass over their weigh-bridge.

I was exceedingly fortunate in meeting a man who was a wholesale potato merchant in Glasgow and was also Potato Controller under the Government for the whole of Scotland. Gathering from his conversation that his knowledge of Jersey's agricultural industry was extensive, I asked him how it was possible for them to grow potatoes year after year without changing the crop, and he replied that it could be done simply by renewing the soil with roots.

He further said: "Tenants here in the low-lying ground pay £20 per acre per annum as rent and in the higher parts of the Island about £15 an acre. The potatoes are placed in rows which are only 1ft 1in apart in comparison with 2ft and over in England. On account of the small space between the rows, this has to be done by hand, but the large number of rows gives a yield of 10 tons per acre in comparison with about 6 tons in England and Scotland." The potatoes are taken up three weeks before they finish growing, as it has been found that this three weeks exhausts the soil to a considerable extent. Moreover, taking them up at this time enables them to be put on the market while prices are still high, and a general average price for a whole year's crop is £8 a ton. This gives £80 per acre for the potato crop alone.

Tomatoes are then planted, and an average yield of £130 per acre is obtained from these, making a total income from the two crops of £210 per acre. Grass is then sown and after it has grown to a considerable extent and good roots have been established the whole is ploughed into the land as manure towards the end of the year, to replenish the soil ready for a repeat of the same crops next year.

A ton of chemical manure costing £9 is applied to every acre, and I suggested that it was this, combined with the mildness of the climate, and the production of the same crop year after year that was responsible for the extraordinary results obtained. My friend said: "No, I myself plant 600 acres of potatoes in Scotland every year, without deterioration of yield, and that has gone on for a great number of years. It is entirely due to the renovation of the soil by ploughing in roots." I further suggested that one could not grow tomatoes in Scotland as a second crop, and he agreed, adding; "But we can grow turnips and carrots instead." The conditions under which people live in Jersey makes one envious. The income tax is only 4d in the £ (or was then!) Rateable values and rates are also very low, and Customs and Excise cheap. A bottle of whiskey costing 12/- in England costs 3/6 in Jersey. The prices of tobacco are much in the same proportion, and almost all commodities can be bought much more cheaply than in England. I purchased a pair of Sir Herbert Barker's boots in St Helier for 35/-, and was told that the London price for the same type of boot was 45/- This statement I afterwards checked and found to be true.

This is typical of prices generally, so I concluded that Jersey was not a bad place to live in. There was no unemployment there, for as soon as a man became unemployed he was given useful work to do, and was paid at the rate of 25/- a week. This rate of pay kept the wolf from the door, but was not high enough to compete with wages paid in ordinary occupations, to which the man would have every incentive to return. No dole was paid to an idle person.

I have referred to the low income tax rate, but there is another very heavy form of tax in the shape of death duties from which Jersey is exempt. I was acquainted with the large estates where, before the war, with income tax and surtax at 12/- in the £, it was necessary to pay 10/6 per £ of the income as insurance against death duties, so that 2/6 in the £ more than the total income was being absorbed by taxation, and insurance against taxation.

There was, therefore, a very strong inducement for millionaires like Jesse Boot and Sir Thomas Houston to become domiciled in Jersey. I believe the estate of Sir Thomas Houston amounted to about £7,000,000, which was a tempting plum to the Chancellor of the Exchequer, and an English law has forbidden people domiciled in Jersey to reside in England except for a very short time, without becoming liable for death duties. I think the period is two months in the twelve months preceding death.

The Chancellor of the Exchequer though he would have a shot at Sir Thomas Houston's estate. Lady Houston fought him and won her case, but after having done so she made a present to the Chancellor of the Exchequer of one million pounds.

Nell, Moses Kellow's first wife, who accompanied her husband on some of his European tours described in the next chapters (Photograph taken by Elliot & Fry)

15. VISITS TO CONTINENTAL ART GALLERIES

I have made frequent and extensive tours on the Continent, and have travelled by all of the best-known routes. Newhaven to Dieppe, Dover to Calais, Dover to Ostend, Harwich to the Hook of Holland, Esberg to London, Newcastle to Bergen, and on most of those routes several times.

I have visited the Capitals of Norway, Sweden, Denmark, Holland, Belgium, France, Switzerland, Germany and Italy.

As regards picture galleries, I have visited the Louvre with its six miles of galleries, but detailed comments on this vast collection is of course impossible. I may mention, however, an incident that occurred on the occasion of one of my visits.

My wife and I had ascended the stairs leading from one floor to another, and confronting us at the top was a large picture of hell. In the centre of the picture was a mountain peak, with almost unclimbably steep sides, and all around the lower part of this peak flames were issuing from a source down below, and rising to a little more than half way up the pinnacle, the peak being clear.

Poor damned souls who were the inhabitants of the infernal regions were to be seen laboriously climbing up the steep sides endeavouring to reach the peak, but when they were near the summit the devil—a huge figure standing alongside—stuck them on a pitch-fork and threw them back into the flaming pit below. This it is presumed being their fate for all eternity.

It was a gruesome picture, and while we were looking at it the voice of someone who had been standing watching us, said in English: "Now will you be good!" The voice was that of an American tourist.

The statuary of the gallery is on the lower floor, and among many world-famous pieces I noticed the Venus de Milo, agreed by sculptors as representing perfection in the female human form.

I also visited the Palace of Versailles, whose walls and ceilings are covered with fine pictures, almost all of a military character. It is a striking feature how much of French Art seems to be devoted to portraying battles and the glorification of war. One can hardly help pondering how, without distinction of nationality, there are powerful military figures like Napoleon and Hitler, who are able to plunge the world into war, misery and destruction, and what is even more astonishing is the support they get in carrying out their malevolent designs! I have visited the Hague Gallery, in Holland, and purchased a picture of the Monk, the Bull, and another famous painting. At Dresden I saw, in that famous town's picture gallery, the Sistine Madonna, considered, I am told, to be the finest picture in the world. This is allotted a room to itself, and occupies nearly all of the end wall. It is asserted that angels are constantly hovering over this picture, and that they can be seen by the devout.

This gallery also contains the finest collection of Peter Paul Reubens' masterpieces, but to me the wonderful imagination he displayed in the design is marred by the gross figures of the women he used as his models, which were, I believe, his wife and daughters.

At one of the German Galleries I saw a picture of the death-bed of Kaiser Wilhelm I, the grandfather of the ex-Kaiser who ended his life at Doorn. The bed was surrounded by Prince Bismarck, Von Moltke, and other notabilities of his time, but on a table by the beside there was a lamp with a moon globe, and it is this moon globe that I want particularly to refer to. The artist had been most successful in giving it the character of luminosity. It was impossible to look at that picture and not have the strong impression that light was being radiated from the lamp.

I cannot refrain from also mentioning the two mural paintings that I saw in the Campo Sanctus at Pisa. One picture covering almost the entire side of a wall portrayed Earth, Heaven, and Hell. On Earth the wicked were feasting and enjoying themselves, the good were all praying. A little to the left of Earth was a huge Angel, about as big as twenty men, carrying an adult man on his left arm like a baby up to Heaven, which was located just above the Earth.

Meanwhile a little devil was trying to pull the man from the Angel's arm, and take him to Hell, but of course without success. Hell was situated below on the left side, and there the sinners were being tortured by little devils, who were employed in the engaging task of pulling off their arms and legs.

What queer and crude ideas have prevailed in the world! On the opposite wall was a picture with a large number of circles, one drawn outside the other. It was a mystery to the ancients why the moon and the stars didn't fall down on the Earth, so they concluded that they were supported by globes of glass, each globe fitted over the next one, and the different paths of the various heavenly bodies was accounted for by assuming that the globes rotated in different directions or planes.

I have seen the Uffizi Gallery and the Pitti Palace at Florence, joined together by a bridge, the former containing a large number of the old masters, and at its entrance statues of the Laocoon, the Dying Gladiator, and other renowned masterpieces of the sculptors art. The Pitti Palace contains newer paintings, which were more pleasing to me because of their better state of preservation. I believe it is universally agreed that the two galleries at Florence represent the finest art collection in existence. One cannot help wondering how much of this fine collection will be left there by the Germans, whether they leave Italy voluntarily or as a vanquished foe.

I also visited the Vatican, and saw its wonderful collection of pictures and statuary, representing the old Greek gods and heroes, which, together with other art treasures, were taken from the Pagan temples. In connection with the Vatican, I cannot refrain from alluding to the Great Picture which covers the entire end of the Sistine Chapel, which one may regard as the Holy of Holies of the Vatican City. This picture was painted very early in the Christian Era, at a time when artists preferred their human figures to be in the nude, but this was rather against the susceptibilities of the early Christians.

The artist commissioned by the Pope to paint this picture, took as his models men about the Papal Court, and one of these he did in the nude. The subject, being somewhat sensitive, protested to the Pope, and demanded that the artist be ordered to paint clothes over his bare torso. This interference with his work was resented by the artist, but he had to obey the Papal command, and did so in his own fashion. He clothed the man superbly, but stuck a pair of asses' ears to his head, which did not please the subject any more than his nakedness.

The outraged man again appealed to the Pope to have the elongated additions removed from the picture, but His Holiness refused again to interfere, and so, in this great masterpiece, the incongruous asses ears remain for posterity to smile at.

The great Cathedral of St Peter's contains numerous pictures, all of course of the highest order of merit, but a characteristic of these pictures which interests me is that, to preserve them for all time, most of these pictures have been duplicated in mosaic.

The infinite patience and skill necessary to do this is almost unbelievable, marble containing markings and colours resembling the paints of the original pictures having to be discovered and pieced together so perfectly that the joints are not noticeable.

The Vatican itself is not capable of being defended, but is connected by a covered roadway in a viaduct which joins it with a Castle of St Angelo, a giant circular building, built as a tomb to the Emperor Hadrian. The walls and ceilings of every room in this huge Castle are also covered by pictures.

Many churches both within and without Rome also contain famous masterpieces. I may just mention the pictures in the Doge's Palace at Venice, and at the Church of St Marco and other churches in the same city, as also the pictures to be found (or at least were to be seen at these places when I was last there!) in the churches of Milan, and in the famous picture gallery at Naples.

Though not coming under the category of pictures, the Dome of St Marco's is in Mosaic, as are also the walls within, and the amount of labour bestowed upon this work can hardly be appreciated. Nearly every town in Europe has its art gallery of greater or lesser importance. I have seen a large number of them, but they are too numerous to mention in detail.

16. CATHEDRALS, CHURCHES, PALACES AND EXHIBITIONS

While visiting the great Cathedral of Notre Dame in Paris, I was passing up the Central Aisle when something was pushed up against me. I turned around and saw a venerable priest holding a long-handled brush, and it was this curious object which had been pushed against me. I enquired of my guide as to why the priest had acted as he had, and was told: "He was sprinkling you with Holy water!" I remarked the absence of water and my guide hastened to assure me: "Ah! The water is only imaginary Holy Water. He hasn't had any real water for years!" I have visited also the beautiful Cologne Cathedral, remarkable for its tall steeples and delicate flying buttresses; also the Cathedral at Milan, built entirely of marble and capable of holding 40,000 worshippers. This Cathedral is surmounted by numerous decorated pinnacles, and the roof is of low pitch, adapted to the marble slab roof cover, and is flat enough to be walked on. I confess to have done a rather foolhardy thing when on this roof. The highest of the pinnacles had a very winding stone stairway, and I think they were intended for the use of steeple-jacks only. I, however, took the risk of climbing these steps to the very top, which I accomplished fairly easily, but the descent from that dizzy height was very trying.

The roof inside this Cathedral appears to be of artistically arranged timbering, but this appearance is wholly illusory as the timber effect was produced entirely by the painter's brush and skill.

I would like also to mention the Cathedral at Florence, with its baptistry at one end. This baptistry is also known as Giotto's tower, and was claimed by Ruskin as the most faultless piece of architecture in the world. My wife and I saw a baptism taking place. There was no one other than the priest, the mother and her child. At various stages of the ceremony unctions were applied to the infant's forehead, the mother looking on with tense anxiety which, at the conclusion of the ceremony, changed to one of great relief and radiant happiness. I do not remember ever seeing a happier face. No doubt the mother thought that her child had been made safe for the Hereafter! The entrance to this baptistry is by large, heavy bronze doors, having on their outer surface panels, executed in high relief, portraying biblical scenes. Quite appropriately, in the bottom left-hand corner is a panel showing the Garden of Eden, with Adam and Eve in their pristine nudity and innocence, the lion lying down with the lamb, and all the other orthodox furnishings of the renowned garden, but with the addition, oddly enough, of three pyramids! I remarked to my guide that this was a picture representing the first man and woman on Earth, and this he admitted. But when I asked him who had built those pyramids, he eyed me curiously for a minute, then said: "That I cannot tell, but you are the first man who ever asked me that question." In Cathedrals, however, there are none that are even remotely comparable to St Peter's at Rome. The fact that the treasures of the old pagan temples that flourished in Rome prior to the Christian era, had been largely transferred to enrich and beautify the Cathedral will give some idea what a treasure house it is. I have heard it valued at £10,000,000, but a monetary value can give little indication of what it contains, its treasures being irreplaceable once lost.

Among other things I saw the Papal Vestments, magnificent in their design and dignity.

There is a semi-circular colonnade in front of the Cathedral, and it is within this space that the vast congregations assemble on the rare occasions when the Pope appears on the balcony after his election, and other momentous exceptions to his rule of monastic retirement. One would be amply repaid for the expense and trouble of the journey to Rome just to see that glorious and beautiful edifice alone.

There are, however, two points that I would like to allude to in connection with St Peter's.

Within the Cathedral, and standing on a pedestal about 3ft high is a statue of the Saint himself.

On one of his feet, his big toe—or what once represented his big toe—comes very near to the edge of the pedestal. The devout, in passing, made it a practice to kiss this toe with the result that it has become almost completely worn away.

My wife, who was with me, after polishing the toe with her handkerchief also kissed it on the principle that when you are in Rome you must do as the Romans do. An acoustical phenomenon of the building is that if one stands near a pillar and the guide goes to the wall at the side of the Nave and speaks in a low and ordinary tone of voice, each word can be heard as distinctly as though he were at one's side. I know, of course, of a similar effect in St Paul's, London, in the whispering gallery, but the phenomenon there may be accounted for by the gallery being in the dome itself. In St Peter's I could detect nothing in the configuration of the building to account for this peculiar transmission of the voice.

I noticed at one church in Milan the High Altar was a blaze of precious jewels—diamonds, rubies, emeralds, sapphires, etc.—and to protect them against thieves it was enclosed in a strong steel casing, which can only be opened when nine priests, each using a different key, are there together to open it. I was fortunate in seeing the casing unlocked, and thus having the opportunity of seeing the Altar in all its resplendent glory.

Among other sacred things at Milan I saw the Scala Sanctus, or Holy Steps, which were the actual steps leading up to the Judgement Seat of Pontius Pilate (or so it is alleged!). To prevent their becoming worn each step was covered by a thin wooden plank, and it was believed that if a person knelt on each of these steps in succession until the top was reached, touching them only with his knees, and not his hands, and saying a prayer on each step, seven years would be taken off the time he would otherwise have to spend in Purgatory. There was a printed notice to this effect by the side of the steps. A passage was provided at the top of the steps to enable the worshippers to return. The wriggling necessary to perform this odd ascent step by step without touching them with their hands was very peculiar.

In the category of peculiar places of worship, I must mention the Jewish Synagogue at Prague. I do not know the exact age of this old synagogue, but it was certainly over a thousand years. Inside there were several lamps burning, and these it was claimed had remained alight for hundreds of years, being kept constantly trimmed and replenished by the devout Rabbis, the money necessary for this constant expenditure having been left by wealthy Jews to commemorate their existence on earth.

None but men were admitted to the synagogue, on the ground that if women were also present the men would be looking at them instead of attending to their devotions. The walls of the synagogue were some six to seven feet thick, and in this thickness were recesses converging from about five feet wide outside of the wall to about two inches on the inner side, and the women—whose souls, apparently, did not matter half as much as the men's souls—had to stand outside the synagogue, quite unprotected from the weather, and view what they could of the ceremony within through the narrow slit provided, while their lords and masters worshipped in comfort.

Surrounding this Synagogue was a burial ground. After this was first filled, a fresh layer of earth was placed on top, and the dead buried in that until it could hold no more. And so layer after layer of earth had been added in succession, until more than half the synagogue was hidden from view. The headstones, however, were not buried, but were raised when each layer of earth was put on, with the result that the top surface was very crowded with headstones laid very close to each other. I was told there were 12,000 of them.

The remarkable thing about those headstones was that they extended back to a period before surnames were in use, and in lieu of this, each headstone was surmounted by a symbol. For example, a salmon might have been carved at the top of the headstone, that symbol having been adopted by a particular family. Later on, when surnames were used, the symbol became a surname, so that if the successor of the deceased person was called John, his full name became John Salmon. I was told that the majority of the great Jewish families of Prague could be identified in this way.

To complete this aspect of my travels, I would like to add that I have visited the Church of John Calvin, the founder of the Calvinistic Methodist Church, have entered his pulpit and stood where he stood when he was preaching to his congregation.

I will make but brief reference to the various palaces I have seen in every country of Europe. Besides the Palace of Versailles I took a day's excursion to the Palace of Fontainebleau.

I saw the Palace of the late Kaiser at Potsdam, with the remarkable old windmill close to it, intruding incongruously. The old miller had refused point-blank to sell his property to the Kaiser, and when the Emperor pointed out that he could, if he chose, compel him to sell out, the miller replied: "Your Majesty cannot afford to pay for it," the meaning being that for the Kaiser himself to compel a poor man by coercion to part with his property would cost him too much in honour and prestige. The Kaiser agreed with this view, and the old windmill remained to this day. This is a story the Germans like to tell.

I will mention only one other Royal Palace, namely the Palazzio de Venezia in Rome. I explored this pretty thoroughly, as it is from the balcony of this Palace that Mussolini harangues, or used to then, the great concourse of people that can assemble in the open space in front of it, and along the Victor Emmanuel Memorial or Monument.

From 1890 to 1934 I visited practically every one of the great Exhibitions in Britain held at the White City, Olympia, and Castle Bromwich, outside Birmingham, and I have also visited many important Exhibitions on the Continent. At Dusseldorf, where I saw a gun barrel for a naval gun about 60ft long, very accurately bored out and rifled, this remained there as a specimen of the work of Krupps, of Essen, who were then employing 100,000 men.

I visited a smaller Exhibition at Dresden, and the famous Fair at Leipzig, which is also a great university town. I observed that most of the hundreds of students walking the street bore those awful scars on their faces, the result of the student duels. In these duels they put a thick collar around the neck to protect the throat, goggles over their eyes, and then proceed to slash one another with razor-sharp swords. The peculiarity of these duels is that it is not a combat in the ordinary way, as each party desires to be wounded, and the resulting scars visible to all beholders thereafter will prove to the world that he is a brave man. I think this, together with the adulation of the military caste which prevails in Germany is an indication of the inherent cruelty of the German race, which glories in bloodshed.

17. STOCKHOLM, PARIS, BERLIN, ROME

I have alluded to some of the noteworthy things that are to be found in the Capitals of Europe, but I will now deal with some of their characteristics not already described.

Stockholm, the Capital of Sweden and the long channel leading from it out into the Baltic Sea is really an Archipelago, and Stockholm itself a group of islands joined by bridges. The waterways between these islands are at sea-level and from a high elevation from which the whole of the city can be seen a very beautiful view is presented. The tracery of the waterways, combined with the green foliage of the trees lining the streets, and the verdure of the parks, combined with the many noble buildings that are scattered over the city, make it in my opinion the most beautiful Capital in Europe. Copenhagen is also beautiful, and contains many fine buildings and places of interest, and I would like to mention one of these—the Parliament house.

The arrangement inside the house for the members is as different to that in our House of Commons as it well can be. Instead of sitting side by side on benches as our MP's do, each member has a separate seat on the floor, and a little desk in front of him. Members do not address the House from their place, but walk up to a platform in front and deliver their speeches from a rostrum.

The President or Speaker, whatever the equivalent to our Mr Speaker is called, sits at the back of the rostrum, and it was curious to note the free-and-easy manners of the House, to see members walking up and whispering things in the ear of the President and then returning to their seats. There was a general air of leisureliness which seemed to prevail over the whole assembly, and anything like excitement was entirely absent.

On leaving the building, the Commissionaire standing at the door called my attention to four medium-sized figures which had been carved around the entrance. The first showed a man holding his cheek, the second a man holding his ear, the third a man holding the abdomen, and the fourth man was holding his head. The Commissionaire said to me: "Look at these. The first one has a tooth-ache, the second one an ear-ache, the third a belly-ache, and the fourth a headache— all through listening to the dreary speeches inside this House." One of the things that impressed me most in Copenhagen was the number of bicycles in use.

It seemed to me that every man, woman and child, ranging from five to ninety years, rode a bicycle, and in nearly every street there were two bicycle tracks adjoining the pavements.

Suppose I describe the street near the Jurist Hotel where I was staying. This is what you would encounter in crossing the street. A wide pavement, a tar-macadamized bicycle track 9ft wide, an island, a tar-macadamized motorway, an island, a tramway, an island, another tramway for traffic in the opposite direction, an island, a tar-macadamized motorway, an island, a 9ft wide bicycle track, a wide pavement. I estimated the total width of the street to be 100ft.

The bicyclists are so expert that when a cavalcade of them are advancing along the bicycle track, they will stop almost as one man when a policeman holds up his hand at a crossing. How they balance these almost stationary cycles is a puzzle to me still. When the traffic was resumed at the crossing all the bicyclists seemed to move forward in one body. All over Copenhagen there was provision for leaving and locking up bicycles in the streets, corresponding to our cab ranks. There would often be as many as a hundred cycles in a line, placed at an angle so that they could be easily withdrawn or locked to a fixture, the owner of each bicycle having his own lock and key.

When I visited Paris I was impressed by the plan on which the city had been laid out. The great Arc de Triomph, erected to commemorate Napoleon's victories, and covered with carvings of military subjects, stands on a high elevation, and radiating from it like the spokes of a wheel are the fine Boulevards, chief among which is the famous Champs Élysées, which is four miles in length, and faces the open archway of the Arc de Triomph.

Beyond this on the other side is the great park, the Bois de Boulogne, and the roadway which come up the Champs Élysées continues in a dead straight line through the Avenue de la Grande Armée for a distance of six miles without deviation.

One word about the shops of Paris. I think they are over-rated, especially their big magazins, and do not compare at all favourably with the shops of London. The only exception is the magazin of Lafayette, which has its counterpart in Regent Street, London. The streets of Paris are far more dangerous for pedestrians than the streets of London, and if there is any rule of the road for vehicles it is more honoured in the breach than in the observance. A drive in a taxi is a unique experience. The driver turns his taxi in any direction in which he can see an opening, and a pedestrian attempting to cross the street can never depend upon which way an oncoming vehicle is going to pass.

In Berlin the Unter den Linden is the principal thoroughfare and competes with the Champs Elysees in Paris and Prince's Street, Edinburgh, for the honour of being the finest street in Europe. At its entrance is the Brandenburg Gate, guarded by soldiers on sentry duty, presumably to emphasise its importance.

It is no less than 198ft wide, and about three miles long, and at the opposite end to the Brandenburg Gate stands the statue of Frederick the Great, and nearby are grouped the main public buildings of Berlin comprising the University, the New Cathedral, 360ft high, the Palace of Wilhelm I, the Opera House, the Palace of Frederick the Great, the National Picture Gallery, and the Museum.

When I visited the Reichstag—then of course still in existence—it was a large building, standing a little to the north of the Brandenburg Gate, built in the florid Italian renaissance style, and at each of its four corners were towers surmounted by pillars typifying German industries. The central part of the roof was covered by a huge glass dome, and the facade consisted of a portico supported by six columns, a pediment showing on its front German industries protected by Germanic warriors, the pediment itself surmounted by a gigantic figure of Germania. This was the building which was destroyed early in the Hitlerite period by incendiarists who were probably in the pay of the Nazis, as a symbol that Democratic power or Government by the people's representatives had ceased to exist in Germany.

During one of my earlier visits to Berlin, I was taken for a drive about the City by the Manager of the Allgemeine Elektrizitäts-Gesellschaft, and among the noteworthy things he showed me was a very wide, fine street in which the fronts of all the houses on both sides was one mass of flowers. He asked my opinion of the display, and I replied that it was a beautiful street.

"Well," he said, "it is entirely occupied by workmen. Each house has a balcony of its own on each floor. The larger houses are divided into flats of three or four rooms, and it is a condition of the tenancies that all the balconies are to be kept with flowers, as you have seen for yourself." On the Saturday morning before Easter, 1914, I visited the great locomotive and wagon works of Orenstein and Arthur Kappel. They were at Spandau, which is the German Woolwich, and just outside the Berlin area.

I invited the Engineer who had conducted me round the works to dine with me at mid-day at my hotel—the Belle Vue Hotel in the Potsdammer Platz. During dinner he asked me what I intended doing during the afternoon, and I told him I was going to Potsdam, which I had not yet seen, although I had already visited Berlin several times.

He offered to keep me company and show me around Potsdam, and while we were there he remarked: "I received a very intriguing letter this morning from a friend of mine in the Army.

He tells me that Germany is massing troops on the Russian frontier, and that everything is being carried out with the greatest secrecy, and that the newspaper had been forbidden from even mentioning troop movements of any kind." Now there is no doubt that this was one of the earlier moves of the Great War of 1914-18.

Germany had decided to invade France and was securing her back door against an attack by Russia, who was France's ally. It will be noted that this frontier move in Russia took place before the murder of the Archduke which was supposed to have been the cause of the War. I wonder if our intelligence department knew what was going on! I will continue the story, as it is interesting. From Berlin I went to Stettin, at the south of the Baltic, saw what was worth seeing there, and then went to Hamburg, where, after some sightseeing, I took a steamer-trip round the Harbour.

We passed very close to the two great ships, the *Vaderland* and the *Imperator*, and it was these two fine ships that were ceded to the Allies at the end of the war, and re-named.

From Hamburg I went to Kiel, principally to see the Fleet. As it was some distance out from the Harbour I took a tram which ran along the coast for seven miles, and found that I was then directly opposite the Fleet. I alighted near a steam-boat jetty and boarded a small steamer on its way back to Kiel.

I sailed right through the German Fleet, passing close enough to some of the vessels to distinguish the sailors features, and see them elevating the guns and doing other practices connected with their duty.

As I had an excellent pair of field-glasses with me, I was able to see the ships that were farthest off as well, an impressive sight. On my return to Kiel I went to a stationer's and purchased a coloured album showing all the principal ships of the German Navy, and among them was the *Goeben*, the *Breslau*, and the *Blücher*, which came into the news at the very outbreak of the 1914 war.

Now to return to Berlin, on that Easter Monday (following the Saturday already referred to) of 1914, I went to Ruhleben in the company of an American who was staying at the same hotel as myself, and spent a day there. A great race meeting was in progress, and I saw all the chief events. The betting was done almost entirely by the totalizer, which gives a square deal but deducts 10% from all winnings. How much poorer the racegoers are as a body at the end of the day may be judged if each person bets on four or five races, and has 10% taken off each time. My special reason for mentioning Ruhleben is that it later became a notorious camp for the British prisoners of war between 1914-18.

The next capital I would like to deal with is Rome, but only to the extent of mentioning some of the unusual things not covered by what I have already written. A striking feature of the Campagna—the marshy belt which surrounds the City—is the very long aqueduct which conveys water from the distant hills to the capital. The famous Appian Way also crosses the Campagna. It is paved in crazy pattern with marble slabs, and was the road by which the Roman legions returned after their world-wide conquests.

Appropriately, at the Rome end of this Appian Way, it passes through a Triumphal Arch, with the name of Septimus Severus inscribed on it, the design of which seems to have formed the model for the Triumphal Arches throughout Europe. The Arc de Triomphe of Paris, and our own Marble Arch in Hyde Park are good examples.

From the height of the Campidoglio one can view the ruins of ancient pagan temples, and among them the facade of the Temple of Castor and Pollux. The story about these is curious. In Pagan lore the Chief god was Jupiter, who was a rather flighty old bird, and had affairs with human women, among whom was Leda, who became pregnant in consequence. This was an awkward complication for Jupiter, as his wife the goddess Juno, was very jealous by temperament.

Jupiter, being a god, could do all the things that men could not do, so he turned Leda into a swan, and thought that he had solved the problem, but not so. The swan Leda laid two eggs and sat on them, and when the eggs hatched out, instead of two cygnets out popped two little boys, one of which was named Castor, and the other Pollux.

The human beings of that period apparently got to know the whole story, and concluded that the two little boys, although not fully fledged gods, were at least half-gods, and as such had to be worshipped. So a Temple was built for them, and at the sides of the facade of this temple, still to be seen, are two high pillars upon each of which are mounted a little boy, one being Castor and the other Pollux, and on the head of each is half an egg-shell to denote the strange way they had entered the world.

Now the odd thing is that when this event took place, Rome had a high civilisation and culture, and is it not a commentary even of present day humans that they will believe the most ridiculous of things if anything of the order of sanctity or the supernatural can be thrown around it.

A little beyond where the Triumphal Arch is now situated, stands the Altar upon which the sacred flame was kept continuously burning, and it was a rule that only the purest and best maidens, drawn from the aristocracy of Rome, were good enough to attend to this sacred source of light and heat. These were the Vestal Virgins, and the house of the Vestal Virgins is close and just beyond the Altar.

There are tablets all around extolling the virtues of individual members, especially laying stress upon their modesty and chastity. The next day after I had seen these tablets, I climbed to the top of the Palatine Hill, and entered the Palace of the Caesars.

My guide pointed to an entrance which led to a secret tunnel connecting the Palace to the house of the of the Vestal Virgins, and I came to the conclusion that human nature has changed but little during the past 2000 years.

Beyond the Palace of the Caesars lie the ruins of the Coliseum. In the centre of one side was the seat of the Emperor and others near for those of the Royal Court, while to their left were those allotted to the Vestal Virgins, who thus held a place of high honour. All around the vast amphitheatre were seats for the ordinary spectators, while down below was the Arena in which gladiatorial combats and other amusements beloved by the Romans took place.

The gladiators generally fought to the death, and when the victor had got his opponent prostrate on the ground, he would put his foot on the breast of the conquered man and raise his sword. If the Vestal Virgins turned their thumbs upwards, the life of the conquered gladiator was spared, but if the thumbs were pointed downwards the defeated man was beheaded. These gentle maidens generally turned their thumbs downwards.

It was in this arena that the early Christians were torn to pieces and devoured by hungry lions. The Romans, not contented with the entertainment provided for them, added to it by having delicious meals served to them where they were sitting; but as the duration of the meal was very much less than the duration of the entertainment, they had a pleasant habit of opening their mouths at the conclusion of a meal so that slaves in attendance could tickle their throats with a feather, thus causing them to vomit the food just eaten. They were then able, immediately afterwards, to enjoy another meal. Ducts were provided to carry away what was vomited, and I personally saw some of these ducts arranged in the front part of the seats at a convenient height.

In the Castle of St Angelo there is a remarkable room high up in the building, which was used as an Audience Chamber by the Popes. The Papal seat faced the doorway, and midway between the doorway and the seat was a hole about a yard square. Below this hole, at a depth of 70 or 80ft rocks could be seen, washed by the waters of the Tiber.

A man obnoxious to the Pope would be invited to meet him in this chamber. A carpet was spread over the hole, hiding it completely, so that when the unsuspecting man was invited to advance towards the Pope's seat he eventually stepped on the carpet and disappeared from view, being precipitated on the rocks below, when his dead body was carried out to sea by the river Tiber. This may seem difficult of credence in connection with a person holding the exalted position of Head of the Christian Church, but in the same Castle of St Angelo I saw the large room in which the Inquisitions were held, and was informed that in the old days the Pope himself presided over these often gruesome and sadistic investigations.

But in those days Heresy was held to be the chief crime of which a man could be capable, because it was antagonistic to the interests of Holy Church, and therefore merited the severest of punishments which ended, as a rule, by the victim being burned at the stake.

Now if these inquisitions, presided over by the Pope, could and habitually did prescribe burning at the stake, it can cause no surprise if an alternative and more summary, even humane, method of ridding the Church of its enemies was resorted to.

Some distance out on the Appian Way it is alleged that this is the point from which Christ last ascended into Heaven after His Crucifixion and Resurrection. The early Christians were a naive and practical folk, and realising that Heaven was a long way up knew it would require a strong impetus for the Lord's ascent to be effective. He had to press so hard on the paving stone that His foot sank an inch into it—and there is the imprint of a very beautiful and perfect foot to be seen in that paving stone to this day. A Church has been built alongside the Appian Way at this point, and derives a good revenue from the devout and curious who come to see the impression.

Not far from this place on the Appian Way are the Catacombs which I visited and inspected with interest. I do not advise anyone to enter them except under expert guidance, as in the labyrinth of passages it would be very easy to lose one's way.

A story which I heard in reference to these famous Catacombs bears repeating. An American was doing Europe as one of a party of tourists, and rather disgusted them by his constant assertion that Americans could lick Creation on anything. They decided to retaliate by playing a trick upon him in the Catacombs. They placed an opiate in his drink, and while under its influence he was placed in a coffin with the lid off, on one of the shelves in the Catacombs.

The party then hid themselves nearby, to see what would happen when their victim recovered consciousness. This he did in due course, sat up, had a look round in the dim light which had been provided by the others, and remarked: "Well, I guess I must have died, and this is the Resurrection!" A short pause followed, and then came the triumphant ejaculation: "Gee! And I'm the first!"

After that, his companions gave him up as a bad job, concluding that a man who could boast that he was first up in the Resurrection was incorrigible.

18. VISIT TO VESUVIUS

I will now move on to Vesuvius and matters pertaining thereto. There was a railway six miles long running from Naples to the foot of the mountain, and another railway up the mountain side to very near the summit. The latter, if not the former also, was the property of Thos Cook & Son.

When my wife and I visited it we were the only passengers except the seismologist, who lives in the Observatory about half way up. This volcano was quite different to my conception of it before my visit. All the mountain-side hitherto seen had been of solid rock, and I imagined that Vesuvius was entirely of rock with a crater-hole at the top out of which the lava flowed during eruptions.

Instead of this, I found that a large part of the mountain consisted of loose ash, not unlike the fine ash to be found under a fire grate, and that during eruptions these ashes were not sufficiently stable to hold the lava until the crater was filled, so that the lava was likely to burst out anywhere at the sides of the mountain, flowing in a stream until it solidified, and became hard rock.

When we reached the upper end of the mountain railway, Government guides immediately took possession of us, two to each person, one holding the left, the other the right arm. This precaution was necessary as the uninitiated might easily walk near the edge of the crater and the loose ground give way under his feet, letting him slide down into the pit.

The guides were there to prevent anyone from going any nearer than about six feet to the edge. The crater itself was a rapidly converging, inverted cone, and at its apex it joined a slowly converging pipe. It was exactly the shape of a funnel such as is used for putting paraffin into lamps, and I think it likely that the rapidly converging section was through the loose ashes, and the slowly converging through the hard rock.

The necessity for the precautions for providing the Government guides was proved by the fact that when Thos Cook & Son first made their railway, they placed their haulage drums, which were at the head of it, at what they deemed to be a safe distance from the crater-edge.

But the ground beneath gave way, and the drums were precipitated half way down the crater before they could be rescued, and irretrievably lost. Other drums had to be provided, and placed at a safer distance.

The top of the crater was about a thousand yards across, and strong sulphurous smoke was issuing from it, which was being swirled this way and that by the gusty wind which prevailed. I was anxious to look down into the crater, and to enable me to do so I had to lie on the ground prone on my face. Meanwhile, the two guides were each holding one of my feet, so that if the ground gave way under me, they could pull me back to safety.

One glance at that sinister, impressive abyss made me realise to the full where the conception of hell had come from. There was the bottomless pit, there was the fiery lava, and there was the sulphurous smoke and brimstone, which would nearly choke one if only the merest whiff was to pass one's nostrils.

To anyone knowing the conditions of a volcano, I can easily understand what a deterrent the threat of having to spend eternity in it would be, and it is doubtless one of the most effective, or at least was during the nineteenth century, weapons in the armoury of preachers.

I once heard a man who was not a church-goer say: "I always subscribe to the Church, but on the same principle as I subscribe to the police. It helps to keep people in order. If you didn't threaten a lot of these fellows with hell-fire, you could never keep them in order." I believe John Wesley was called the apostle of hell-fire. I cannot refrain from making a digression to tell a story of something that happened in North Cornwall, near the tin mines. On account of the poverty of the Churches, the pulpits of ten had to be filled by amateur preachers, who were uneducated men. One of these was holding forth to his congregation on the awful terrors of hell, and said: "Hell fire is tremendously hot. I cannot adequately give you an idea how hot it is, but you know the terrible heat of the furnace where they smelt tin around here—well, that's ice-cream to it." Just below Vesuvius is Pompeii, which was completely buried by ashes during an eruption.

These ashes have been cleared away, to disclose how a town was planned in the olden times.

The streets were paved, and were only wide enough for a single vehicle to traverse. There was a narrow pavement on each side, but no channels, so that the streets must have become very slushy. Stepping stones provided for pedestrians' use in cross-roads testify to this fact.

I observed in many of the houses dados of a waxy-looking substance, on which were painted deer drawing vehicles or disporting themselves in graceful convolutions. The result was rather attractive.

At the entrance of one of the doorways I saw the figure of a fierce-looking dog, and the words "Canus Canem!" (Beware of the Dog!) all done in mosaic. Almost everything of interest in Pompeii has been removed to a museum in Naples, which I visited. I am afraid that the old Pompeians were guilty of a good many things that are not regarded today as admissible in polite society, and the result is that many of the exhibits are contained in private rooms in the museum, to which men only are admitted. When I was there I found a crowd of angry women outside these rooms, protesting against their exclusion and demanding to know why, if men were admitted, they should not also be allowed inside. Well, women's rights had not extended that far when I was in Naples!

19. THE SWISS AND ITALIAN LAKES

The first of the Swiss and Italian lakes which I saw was Zurich. I was staying at an Hotel by the lakeside, where in the evening the guests were wont to sit outside, enjoying the warm summer weather. It was there I had my first real café Malanche, not the poor imitation of it we get in this country.

While I was enjoying my siesta, a man seated a yard or so from me, suddenly jumped from his chair and began running as fast as his legs would carry him. It turned out that he was a thief who had mixed with the hotel guests, and had tried to pick a man's pocket.

The other guests immediately set chase, and to avoid capture the man jumped into the lake, and tried to swim to the other side with his clothes on. These so hampered him that it was a matter of speculation among us as to whether he would be able to reach the farther shore.

However, he did succeed, but only to be taken into custody by the gendarmes, who were waiting to receive him. I walked round the end of the lake and witnessed the arrest.

While in Zurich I noticed from my bed, at 1 o'clock in the morning, a light shining on my bedroom window, and got up to investigate. I found it came from a house which was burning furiously half a mile away. I went back to bed and to sleep, but a full hour and a half later I was again roused, this time by the tramping of feet, as though a regiment of soldiers were passing by. Again I went to the window, and saw a sedate body of men marching in true military style in the street below. Hearing a sound in the corridor, I opened my bedroom door and enquired of a manservant outside, what this military parade signified in the middle of the night. The reply I got was: "Oh, those are the firemen, sir, marching to the fire!" I had been accustomed in London to seeing galloping horses, and later fast motored engines, flying at breakneck speed, with firemen fully equipped standing on either side, ready for instant action, while the traffic gave them right of way. The leisurely method of dealing with a serious fire which prevailed at Zurich struck me very forcibly, by contrast with our swifter method.

From Zurich I went over the top of the Rigi and down on the other side to a little village called Vishnau, which is on the side of the Lake of Lucerne. Lucerne is rightly the Mecca of all tourists who delight in natural beauty. The mountains surrounding the lake are all of a high elevation, and in the clear air of spring distances are so much fore-shortened that while one is basking in the hot sun, at the level of the town and the lake, the snow line on the high mountains looks almost within a stone's throw. The dazzling whiteness of the virgin snow, suspended in the limpid air between the greenness of the vegetation and the intense blue of the sky above, reflected with equal intensity in the calm lake beneath, make a vivid combination of colour that is deeply impressive.

A little way up the hill facing the lake, is an hotel—the Minerva—at which I have stayed on several occasions. In lieu of walls the whole of a very large dining room is glazed with plate glass, and looking out over the lake flanked by the Pilatus, the Stauzerhorn, and other noble mountains, I do not think that the panorama seen from the hotel windows can be excelled for beauty anywhere in the world.

I spent a night on the summit of the Stauzerhorn in order to witness the sunrise over the Alps at about 4am From this peak it seemed but a stone's throw to the summit of the Pilatus, so clear was the air. The actual distance was seven miles.

Of course, it is the natural scenery which is the great attraction at Lucerne, but there are two other features in the town to which I will allude. One is the quaint old bridge over an arm of the lake, which is roofed over and contains paintings on the underside of the roof, and the other is the museum of Peace and War, in which it is stated that there are samples of arms from every country in the world, except one—the exception being Great Britain.

They refused to show anything from Great Britain on the ground that Great Britain is the most aggressive of all the nations, and they express their disapproval by thus refusing to allow anything from Great Britain to appear in their museum. In view of this gross insult to our country, I do not think that the inhabitants of Lucerne deserve the extensive patronage they get from the British people.

I travelled from Lucerne to the town and lake of Lugano by steamer along Lucerne Lake to the other end, where the village of Fluelen is situated, and then by a rapidly rising railway to the mountains beyond. It was interesting to look from these heights into the deep valleys below, and see the rushing torrent of the river Reuss, which supplies the lake of Lucerne.

In view of the really hot weather and cloudless skies this flood of water seemed inconsistent, but the rivers of this part of the world are always torrential in summer because of the melting snows.

Shortly after leaving Fluelen a very interesting state of things takes place. As I was travelling along the railway on the hillside, I noticed a little church about fifty yards farther up the hill, the train then entered a tunnel through which it travelled for apparently some miles.

When it emerged, I happened to look out of the carriage window on my left hand, and noticed a church about fifty yards below the railway track. The train then entered another tunnel, somewhat similar to the first, and when it emerged from this second tunnel, I again noticed a little church down below, but now it was about a hundred and fifty yards below. I recollect thinking how similar these little Swiss churches were to one another, but what had actually taken place was this: The tunnel we had first entered, instead of going in a straight line, had taken a spiral course inside the mountain, and after completing a circle had again emerged at a point directly above the entrance below. The same thing had happened in the second tunnel, so that it was in reality the same church I had seen each time, though of course it appeared lower down. This was the method adopted by the railway engineers to gain height where a straight line would have made too steep a gradient.

The town of Lugano contains many fine hotels and is definitely a tourist resort. In a church near the lakeside is a mural painting of the "Lord's Supper" by Leonardo da Vinci which is probably the most copied picture in all Christendom.

A short distance from the town of Lugano the frontier between Italy and Switzerland crosses the lake, I took a steamer from the Swiss to the Italian side of the lake, and there ascended by railway the mountain of Monte Generoso.

From its peak a magnificent view of the surrounding country can be obtained. When looking at picture postcards, I had always regarded them as being too highly coloured, but from the top of Mount Generoso I saw an intensity of colour all round me that vied with that of any picture postcard I had ever seen The mountain sides are everywhere steep, and looking down upon the waters of the lakes below, their colour appeared to be of a deep indigo. The greens in their variegated shades were of an equal intensity, while the colours of the different rock strata exposed by erosion and ancient upheavals were equally pronounced. This was no doubt due to the extreme brightness of the sun and the clearness of the atmosphere at these heights. But it was from the peak of Monte Generoso more than any other place I recollect visiting that this extreme intensity of colouring was observable.

The lake of Como lies at the foot of the town of that name, and what struck me when I first saw this stretch of water, was the large number of washer-women who were busy with their laundry at the lake-side. With my back to the town, I turned along the left bank of the lake and came to an interesting little village about half way along its length, and from there was rowed across to the other side in a small boat, to where the famous Villa Carlotta is situated.

This villa is not only a treasury of art in the way of paintings and sculpture, but it was claimed that the extensive grounds contained at least one specimen of every known plant and tree in the world. This, of course, was an exaggeration, but it did contain a sufficient variety to make it interesting.

After a day's trip to Lake Maggiore, I returned to Lugano, and from there went to Stresa—where one cannot help being impressed by its amazing display of flowers. They are encountered everywhere in profusion.

The beautiful hotel at which I was staying near the lake side had one very large room literally full of flowers, and the name of the hotel appeared in a large floral design outside.

There was a good deal of bathing, including facilities for high diving, but unquestionably the feature of the lake was its four islands, of which the Isola Bella is the largest and most beautiful. It contains a Royal Palace and the beautiful way in which the whole island is laid out in trees, shrubs and flower beds makes it one of the loveliest spots in Europe. It was on this island that the unfruitful Stresa conference was held.

While on the subject of lakes, I would like to refer briefly to lake Geneva, with the town of that name at one end and Montreux at the other. Montreux rather than Geneva is the favourite tourist resort, and contains a large number of hotels. The Palace Hotel is a large and imposing building, and it is (or was) a fashion in Montreux that, at whatever hotel one might be staying, it was "the thing" to go to the Palace Hotel for afternoon tea.

At Geneva one gets an excellent view of Mont Blanc, and the town was, of course, the home of the League of Nations. What may now develop out of the League of Nations, if anything, is in the lap of the gods.

An amusing little episode took place during my journey from Montreux to Geneva. Shortly after leaving the latter town I asked some of my fellow passengers if they had ever seen rose-coloured swans on the lake of Geneva. They answered no, and asked where they might be seen.

I replied that they were at the Montreux end of the lake, and I would show them the swans in due course.

The story spread throughout the steamer, and when we were approaching Montreux I found myself surrounded by a large and eager crowd. Now at this end of the lake there are usually a large number of ordinary white swans, and when we reached these, I pointed towards them and cried excitedly: "Look, there they are!" "But those are just white swans!" was the disappointed comment. "You said 'rose-coloured' swans!" "Well, aren't they just the colour of beautiful white roses?" I asked. They were very merciful to me, and it was decided not to drop me in the lake.

The great mountain near Montreux is the Rocher de Naye, which can be ascended by a mountain railway, either from Montreux itself of from a little village called Territet. About half-way up the mountain this railway passes through a tunnel, emerging beyond into a basin-shaped hollow, where tinkling sounds at once strike the ear. Herds of cows are put here to graze, and the sound comes from the numerous cow bells, which are a necessity at these lonely heights, where cows are liable to stray, and the only guide an owner has of any of his missing herd is the sound made by the bells.

The railway ascends far above the snow-line, and I presume is unusable except in summer. It was interesting to wind in and out around the snowy peaks, looking down the steep slopes to see trees growing out of the snow. It was thus I crossed the Bernese Oberland, passed through Berne and on back to Lucerne. I had thus travelled to Lucerne by three different routes—by the one just described, over the Rigi from Zurich, and by rail from Basle.

20. VENICE

Although one has been prepared for the unexpected in Venice, I shall never forget my sensation when I emerged from that city's quite ordinary railway station, to find, instead of the usual row of taxis to take one to one's hotel, a long line of shallow steps leading down to an expanse of water—the famous Grand Canal, where gondolas waited to take one wherever was desired. I stepped into the nearest of these peculiar mediaeval craft, and gave the name of my hotel.

The Grand Canal is shaped like the letter S, but there are cross channels which can be traversed to make short cuts to any part of the City. My hotel—the Grand Hotel de Venise—was at the mouth of the Grand Canal, where it debouches into the sea, and a fairly long trip from the railway station.

The gondolier stood at the stern to row his craft along, and after about a third of the journey was made in almost complete darkness, a strange voice suddenly addressed me in Italian. It sounded disturbingly close to my ear, and I found that there was a man who had apparently nothing to do with the gondolier, lying full length behind me, with his head close to mine. I think his intention was merely to indicate to me the points of interest revealed by the clustering lights we passed, in the expectation of the inevitable tip, but the fellow was so repulsive in appearance, and his manner so insinuating that the only tip I felt like giving him was a tip into the not-so-limpid waters of the Canal.

The Grand Hotel was not only delightfully situated, but meals, especially breakfast, were served under copious umbrellas on an open terrace by the waters of the Grand Canal. The walls inside the hotel were decorated with original mural paintings of quite decent merit. It was also situated in the most interesting part of Venice.

Close by was the famous Piazza, a large open rectangular space, with St Marco's Cathedral at the extreme end of it, and next to it the equally renowned Doge's Palace, which once housed the Doges or Dukes of mediaeval times.

This palace is very extensive, and is best seen with the aid of a guide. It contains a large hall where Justice was dispensed, and leading from it over a narrow canal is the famous Bridge of Sighs leading to the dungeons. Many of the unfortunates who crossed over this bridge saw the light of day for the last time through its crevices, as most of those sent to the dungeons were kept there in total darkness until they expired.

About half way along the Grand Canal is the Rialto Bridge, with shops on each side, much as I believe the old London Bridge used to be. In line with the front of St Marco is a crooked street, so narrow that the tips of one's fingers, with one's arms outstretched, can almost touch the houses on either side. This crooked street leads to the Rialto Bridge, and it is in this street and the area surrounding it that the really quaint old-fashioned shops of Venice are to be found.

All along the Grand Canal are the fine houses of the great Venetian Princes of the olden times, and it is remarkable what a lack of originality there is in our modern architecture that instead of developing new forms our modern architects go back to the Grand Canal of Venice and to Rome and Athens, to get their inspiration.

Most of our great buildings in this country are built on the model of Greek or Roman temples, or the houses of great Venetian merchants. Slavish copies abound of the two former types, with their Doric, Ionic, or Corinthian capitals, their architraves, entablatures and cornices, combined with the low-pitched roofs adapted for a covering of marble slabs.

As in the original design shade was the chief necessity, the spaces between the columns were not built up, but in our colder climate walls have to be built, which would be the natural support for the roof, and the columns are a redundancy, except as ornament, But as utility, or at least apparent utility, is the first essential of beauty, it is doubtful whether the retention of this form of architecture, which is unsuited to our climatic conditions, is consonant with the principles of true art.

As matters stand we cannot design a Royal Exchange, or even a humble book-case, without going back 2000 years for our inspiration. At the same time, I think the pendulum has swung too far in the other direction in certain cases, where steel and concrete govern the design to a certain flatness over extensive areas, making it unduly monotonous.

The holiday resort of the Venetians is the Lido, where there is plenty of sunshine as well as sea bathing. One evening, returning from the Lido to Venice, I saw quite the most remarkable sunset that I had ever watched. The horizon was fringed with thick clouds of inky blackness, standing up in definite forms like castles or mountains built up of solid substance. Above this black silhouette shimmered a pale green, luminous sky through which shot swathes of a deep, fiery red emanating from the sun which was behind the black nimbus clouds.

Gazing at the magnificent scene, I asked myself whether, if it were transferred to canvas by a master artist, it would be ever admitted as a portrayal of a natural phenomenon. I concluded that it would not.

Proceeding towards Venice the water was studded with boats with all their sails painted green and russet, and as the darkness deepened, the dancing lights from these boats, with the illuminations of Venice and the Grand Canal in the distance shivered by the lapping waters, combined to make a panorama of a unique and beautiful character.

Another feature of the sea-City that I found attractive was the custom of listening to the gondoliers singing to audiences in other gondolas, in the cool of the evening. The velvet darkness, the limpid stars, the charming Italian voices raised in song have remained a vivid recollection to this day. One paid, of course, for these entertainments.

There was for me, however, one very definite "fly in the amber" in connection with Venice, and that was the plague of mosquitoes. Sleep at night would have been altogether impossible were it not for the mosquito netting that surrounded my bed. Even so, notwithstanding every precaution, some mosquitoes managed to get inside, and soon after the light was turned off I would hear that shrill, buzzing note which gave warning that the infernal insect was after my blood.

It was useless to ignore it, and I had to switch on the light again and search diligently for a tiny speck in the white netting. I would kill it and settle down again, but the chances were that within two minutes of switching off the light that confounded buzzing would recommence, and the tiresome search would have to be done all over again. I got barely three hours sleep on any night I stayed in Venice, and became badly bitten. Indeed, nine days after leaving Venice, when I was at Morano, I was a mass of itching weals from head to foot as a result of the bites that I had sustained from the venomous insects of Venice. So intolerable was my misery that I had to consult a doctor, and obtain appropriate treatment, which, fortunately, was successful. I concluded that the Venetians must be a tough people. They need to be—they have the Germans on their hands now! But not for long, let us hope!

21. THE ALPS

I left Venice for Cortina, travelling alongside the river Piave. I made many excursions from Cortina d'Arpezzio, which is in the heart of the Dolomites. From there I travelled through very remarkable country to Bolzano, where a railway comes from Innsbruck, intersecting the Dolomites in a line at about right angles to the direction I had entered Bolzano from Cortinal d'Arpezzio.

I had previously travelled from Innsbruck to Bolzano by this railway. The Dolomites being of magnesia limestone, the rock was soluble in water, and as water was found mostly flowing at the bottoms of valleys, these became deeper and deeper relative to the mountains on either side. The consequence was that, in travelling the mountain roads of this part of the Alps, one had frequent glimpses of valleys of unbelievable depth.

The shapes of the mountains was the result of the solubility of the rocks, and sometimes produced fantastic results. I saw one mountain with five extremely sharp fingers standing very close together, and looking very like a giant's hand pointing upwards.

From Bolzano I continued my journey to Merano, known as the Garden of Europe. I stayed at a magnificent hotel, whose entrance would have done credit to any city. The rooms were spacious and well furnished, the meals excellently cooked. The hotel was surrounded by beautiful grounds, yet the charge was only 8/- a day of our money, due to the then existing rate of exchange, it being 104 Lira to the £, instead of the pre-war 25. I am sure, though, that at the same time in Switzerland, where the rate of exchange was 23 francs to the £ in comparison with a pre-war 25, the charge for equivalent accommodation would have been between £2 and £3 a day.

Friends of mine, whom I will mention later, stayed at Merano shortly afterwards, in another hotel perhaps not quite so luxurious, and the charge, including all meals except lunch, was only 4/- a day.

From Merano I went by road via Sponidig to Trafoi, which is a village at the foot of the Great Otler—13,000ft high, and is the starting place for the Stelvio Pass. The road at the top of this pass is the highest in Europe, being 9,200ft above sea-level .The climb is by a zigzag road, which does not inspire one with its safety. Each turning is so sharp that no motor vehicle can get around it without stopping and backing, and then re-advancing a little more round the bend.

The road is not fenced in any way, and when the car is being backed to the edge of what is virtually a precipice, and one has to rely on the care and skill of an unknown driver, one begins to think of one's sins, and speculate upon one's chances in the Hereafter. However, I eventually reached the summit in good order, and found the ground strewn with spent cartridges—the aftermath of an engagement during the Great War.

From here another zigzag road took us downwards, but the bends were not so sharp as on the ascending side. Near the bottom we turned into what I presumed was an ancient hostelry with an open courtyard, and exceedingly rough accommodation within. Our driver, dressed in his sheepskin coat found congenial companionship in a room full of his compatriots, and I surmise dined and wined to his complete satisfaction. As for us passengers, we had to be content with a very mediocre glass of wine, and being teetotal myself, I remained thirsty. But it was not long before we arrived at Tirano, the starting point for the Bernina railway, where accommodation was more civilised.

Next morning we ascended the Bernina Pass on a spiral of railways inside the mountain in the manner already described on our journey from Lucerne to Lugano. In order to get the best view I left the train one station before the top of the pass was reached, had refreshment at a restaurant, and walked right over the Pass to the next station of the other, descending side. I thus had a splendid view of the country on both sides of this famous Pass.

The journey from here to Pontresina, our next stopping place, can only be described as superb. We passed several glaciers on our way, and at one point the valley is crossed by a tremendous viaduct of astonishing height. Looked at from a distance, it seemed a frail structure to carry a locomotive and train of carriages, but when I saw a train passing over its lofty track, the train seemed but a toy itself in comparison with the immense series of archways.

Making Pontresina our headquarters we explored the country around, and in particular visited St Moritz, famous for its winter sports.

I stayed some time at Grindelwald, where I was surprised to find in the hotel in which I stayed, that the bedroom floor was made of marble slabs, which, though not uncomfortable during the warm weather when I was there, would, I surmise, be very cold in winter.

This hotel faced three great mountains, the Monck, the Eiger, and the Jungfrau. I climbed the Monck, which had very steep sides, and was drawn up a thousand feet on a rope. I then joined a roped party to explore the huge glacier which lies between the Monck and the Eiger.

This was a risky business, as the glacier was traversed by crevasses at least fifty feet deep, often occurring not more than six feet apart. We had to walk warily between them.

Though these crevasses were more or less parallel, this was not always the case, and I have a vivid recollection of having to jump a cross-crevasse which joined two parallel ones. The width I had to leap represented the limit of my jumping powers, and bearing in mind that it was all ice underfoot, and a slip on either side would have meant a fall into the crevasse, my feelings at the time can be better imagined than described. To make matters worse, I had the fear that, were I to slip, I would inevitably drag some of my companions with me, and not being accoutred for Alpine climbing I afterwards judged that it was a foolhardy thing to venture on such a project unprepared.

I went up the Jungfrau as far as the Great Shiedig, which is the limit that ordinary tourists go to, as from there to the summit was considered only fit for experienced climbers to venture.

There was an interesting ice grotto at the Great Shiedig. Looking across the wide valley at the foot of the Jungfrau, a snow cap a hundred feet thick could be seen on the mountains opposite, and many times during my short stay there, I saw huge masses of snow break away from the snow cap, and rush down the mountain side. There were numerous ravines gashing the lower mountain slopes, and these avalanches of snow found their way into them like water finds its way into a river bed. For a considerable time after the avalanche had started the ravine would appear to be full of white, tumbling water, and no one, except he had seen the avalanche start, and noticed that the flow was intermittent, would have thought otherwise.

It was highly dangerous to cross these ravines in the intervals of the avalanches, as one might come without warning, and bury the too-venturesome traveller. Nevertheless, I saw two parties start off from the Great Shiedig accompanied by a Guide, to make one of these perilous journeys.

From Grindelwald I took an excursion through Interlaken to Lauterbrunnen. This valley is flanked on both sides by high mountains. On the left-hand side, looking up-valley, is an exceedingly deep ravine cut by a fast flowing river, but for a considerable distance this river has now disappeared. I climbed high up the mountain in order to take a pathway into this ravine. At the inner end I found a chamber lighted by various coloured electric lamps, and the huge torrents of water representing the river disappearing into a hole in the ground.

Below this point, no water was flowing in the old river bed, but near the bottom end of the ravine the water was again issuing forth with terrific force from a hole about 7ft in diameter.

I believe the water of this disappearing river is known as the Trummelbach, and its source is a glacier on the Jungfrau. On the right hand side of the valley is the Staubach Fall, which has sheer drop of 1000ft, with the result that before the water reaches the bottom, it divides into an exceedingly fine spray, which in the morning sun looks like a silvery veil, and is wafted to and fro in the breeze.

In company with my wife, my little son 10 years of age, and my mother who was 63, I climbed to the top of the St Gotthard Pass, starting from near the mouth of the St Gotthard Tunnel. On our left in the upward journey we passed the Benedictine Hostel or Abbey where the world-famous Benedictine liquor is made, the secret of whose manufacture is supposed to be known only to the monks.

Cars were provided at the foot of the hill to convey those unable or unwilling to do the climb on foot, but my energetic mother scorned the idea that she could not do the walk, and it was with admiration that I observed her herself leading the band of walkers, and despising the laggards.

As we neared the top we passed the Devil's Bridge. The story about this bridge is as follows: The tremendous wash of water passing down the river broke away the foundations of one bridge after another constructed by the inhabitants of the locality, and taking advantage of their predicament, the devil offered to build the bridge himself, providing that he got the first living thing that passed over it.

After his Satanic Majesty had constructed the bridge, a bridal party came along, and the devil's eyes glistened as he saw the beautiful bride he was "to have and to hold" as the lawyers put it. But just in the nick of time a shepherd came along with his dog. The bridal party did not know of the devil's bargain, but the shepherd did, so he whistled to his dog and sent him first across the bridge. So the devil was baulked after all, for his intention had been to snatch a human being, but he had not been careful enough in the wording of his bargain—and got a dog instead. Thus the inhabitants got a sound bridge, very much on the cheap.

At the top of the St Gotthard Pass there is a small mound, and our guide, standing on it, said: "If I had a bucket and a cup, I could throw a cupful of water that way" (pointing to the West) "and it would go into the Rhone. And if I threw another cupful towards the North, it would go into the Rhine. A cupful thrown towards the East would go to the Danube, and another to the South would go to the river Po." This assertion struck me as being very odd, and its meaning perhaps obscure to those unacquainted with the geography of Southern Europe. However, no-one made any comment, and neither will I, leaving it to my readers to judge whether there was any justification for the oddity which appealed to my imagination.

22. CHANCE MEETINGS, MONKEY GLANDS AND PICKPOCKETS

Talking of rivers, I think I may refer here to my experience on the Rhine. I have travelled at least a dozen times by rail on the west bank of the river running upstream from Cologne, and once on the right side of the river backwards and forwards between Frankfurt and Luxemburg.

I have also sailed on the river between Cologne and Mayence, passing the vineyards on the steep bank of the river, which are in terraces supported by retaining walls.

The right bank of the river is very high, and all along the top are castles of the old German barons. At one part there is a remarkable ferry. A boat is moored to a post in the middle of the river by a long rope, and the boat can be made to travel from one side to the other by simply moving the rudder against the flow of the river, this supplying the entire motive power. I have seen the falls of the Rhine at Schaffhausen, both during daylight and at night, when they were illuminated by various coloured lamps projected from great search-lamps.

I went to the South of France via Avignon, Marseilles, and Cannes to Nice. This is a convenient place for excursions to places in the neighbourhood. France is noted for its well planned straight roads which were made by Napoleon for military purposes. Probably the most famous of these is the Grand Corniche which runs along a considerable length of the high ground overlooking the Mediterranean Sea, and gives an excellent view of the coastal scenery of the South of France.

Another road, the Lower Corniche, runs nearly parallel to the coastline and traversed the more densely inhabited parts of the area. Most of the excursions inland are interesting, and one in particular. This goes right into the heart of the country via Grasse, where the highest grade of natural scent is manufactured, and thence to Cannes and back again to Nice.

My wife and I had taken several of the more popular excursions on large charabancs, and on these we had noticed a young couple who seemed to be doing the sight-seeing pretty thoroughly. On our fourth excursion, however, only a small car, holding six persons was used, and two of our fellow passengers were the couple I have mentioned.

When six people are keeping company for a whole day they naturally fraternise, and something in the appearance of the young man suggested that he was Welsh, prompted me to ask: "Cymro ydach chi?" His face lit up at hearing his own tongue, and he answered: "Ie, glan gloew." It turned out that he was a dentist from Bala, and was on his honeymoon trip. We had many mutual friends in Merionethshire, including members of the County Council. He was related to Sir William Preece, whom he claimed was the real inventor of wireless, and not Marconi.

Now the other extraordinary thing was that the two remaining fellow passengers were from Huddersfield, and were intimate with my wife's family there. Thus did we all meet by a happy chance.

From Nice we travelled via Mentone to Genoa in Italy, passing on the way the establishment of Dr Voronoff of monkey-gland fame. We actually saw a cage containing a number of almost human apes, who were to be the victims of his next operations, and I wondered, with their high intelligence, whether they realised at all what lay in store for them! Genoa is the largest port in the South of Europe, and it seemed strange to me to be told that the great showplace of this City is the Campo Sanctus, or Cemetery. There is one long corridor with recesses like rooms on one side of it, in the wall of which the deceased is buried, and in the recess of room between the wall and the corridor are life-sized marble figures of the departed's mourning friends, dressed in the fashions of the period, and posing in various attitudes of grief. In other places the coffins are buried in walls, with brass tablets giving the name, etc., of the deceased occupant. Many large and sometimes elaborate mausoleums dot the rising ground, which forms part of the Cemetery, but I think one of the most unusual features is a large piece of flat ground in the centre, where poor people are buried. Their humble bones have only a limited lease, for these bodies are cleared out every five years so that the ground can be used all over again.

Near the docks are broad streets with high buildings, but at frequent intervals narrow alleys connect the broad streets, and these alleys are used for drying clothes, ropes being placed across, at varying heights, from one block of buildings to the other.

Looking along one of these narrow back streets one sees an extraordinary sight, as from end to end it is almost filled with articles of clothing, and they obscure the light to such an extent that at the street level it is almost dark.

Genoa is noted for the number and skill of its pickpockets. I was told that they go about with cut-throat razors and can, with an almost lightning stroke, cut a pocket which they think contains valuables right out of the garment.

I was myself robbed of twenty-seven pounds, so have little that is good to say of Genoa. I would say that any stranger who attempted to pass from wide street to wide street would possibly put his life, but certainly would place his property, in extreme jeopardy.

23. CASINOS

On our return journey from Genoa to Nice we stopped at Monte Carlo, and visited the Casino, the latter being situated on a hill-top. Play was in progress, and as an interested visitor I came away impressed with the idea of what fools human beings can be.

The owners of the place are better mathematicians than their customers, and they know that the winning number and colours are governed by a very rigid law of mathematical probability, and if the rules of the game are made with a balance of chances in favour of the bank, the inevitable consequence will be that the bank will be the gainer, and the crowds that enter to play will collectively be the losers.

The shallow pates who patronise the tables have a sublime belief in luck, and seem incapable of consecutive reasoning. Put into plain language it means that some supremely powerful being, god or spirit is there controlling the Roulette wheel, causing it to stop at a particular place, so that the favourites may win money, and those out of favour lose.

I saw players with notebooks ruled in two columns headed red and black. They watch to see if red or black comes to the winning point, and then mark it down on the appropriate column. If for instance it stopped on red three times in succession: "Oh!" says the nincompoop, "the luck is running on the red," and he will hasten to put his money on the red before the luck changes.

If it should turn up on the black the fourth time, he consoles himself by saying that he had left it too long, and should have put his money on earlier.

Counters are used instead of money at the tables, and many gamblers believe that by manipulating their counters in a certain way before placing them on, it will bring them luck. I stood behind one lady and watched the performance.

She would take, say, two counters, put them at a particular place on the table, place others in what was evidently a pre-determined pattern, then shift the piles about, sometimes putting one on top of another, and finally placing her counters on the spot where they are in play. If she won, a smile spread all over her face, which meant as clearly as words could say it: "Am I not clever?" Of course, the gaming tables produce an enormous revenue for their owners, because the stakes are high. The town of Monte Carlo is full of luxury hotels, to help empty the pockets of the already empty-headed people who visit the place.

However, I have also seen the over-gilded rooms of the Kursaal at Weisbaden, where chemin-de-fer is played, and also the more proletarian places where Coule is played. In the latter game the chances are preposterously in favour of the bank, but in the one I visited in Montreux the stakes were limited to five francs, which at that time meant about 4s.2d.

In the centre of a very long table covered with green baize was a large bowl about four feet across, and about as shallow in depth as a saucer. Around the periphery of the flat central part were two or three rings of pocket holes. The croupier would take a boule and throw it so that it revolved near the outer edge of the bowl, but as it lost momentum it naturally came nearer towards the centre, finally bumping over the shallow pocket holes until it came to rest in one of them.

These pocket holes were numbered from one to nine and chalked in squares on the green baize. At each end of the table were two rows of figures, one line having the odd numbers 1,3,7,9, and the other the evens 2,4,6,8. The players could bet on the odds or evens, and if they won they were paid two to one.

It will be noted that the number 5 is missing, and when the boule dropped into the unnumbered pocket hole that should have corresponded to 5, the bank scooped in all the money that had been placed by the betters on the table. This meant a 1 in 9 chance in favour of the bank, but there was another way of betting and that was to put the money on a single number, and if the ball settled on that number, the bank paid out 6 to 1. It will be seen that in this case the bank had three chances out of nine in its favour, and that taken collectively one-third of all the money put on the table by betters on single numbers, went to the bank.

I think the most astonishing about these gaming rooms was the way in which the croupiers, who collected and paid out the money, did their job. There was a croupier at each end of the table, players sitting side by side all round. There were also rows of players standing behind those seated, and reaching over them to place their money on their selected numbers or rows, or to collect their winnings.

Apparently, the croupier would be taking very little notice of what the crowd was doing. Yet without a written note he knew exactly what numbers each person had bet on, the amount of his bet, whether he had lost or won, and how much was due to him if he had won, with unerring accuracy.

He had a long-handled shovel, and from near the centre of the table he would push this shovel out to collect the money stacked there, and also to convey to winners the money due to them. The quickness with which each operation of spinning the boule and paying and receiving the cash was done, was scarcely credible, but a boule table of this kind must have been a veritable gold mine.

I believe at Montreux it was run by the Corporation. The hold which the gambling fever can get on a man although the chances are so much against him is surprising. A young man staying at the same hotel as myself went to the gaming tables with me, and seated himself to play. From the opposite side I watched him, and his face became whiter and whiter, until it was like a sheet. Big beads of sweat stood on his forehead and he began to tremble visibly. He seemed in a state of near collapse, when I went to him and dragged him away from the table and accompanied him back to the hotel.

24. TRAVELS IN NORWAY

I am not writing a Guide Book, not attempting to describe things in chronological order or geographic propinquity, but rather to describe unusual incidents that have occurred to me in the course of my travels.

When travelling from Bergen to Oslo in Norway, I saw a girl of about eighteen years of age at a wayside station, and her hair was literally like threads of pure gold, not what is euphemistically called golden, but the colour and appearance of the metal itself. It was the only example of its kind that I ever saw, and it was certainly the most beautiful.

At Amsterdam I saw an apparently respectable lady walking along the street with what looked like a smooth golden skull on her head. The entire pate, from the top of her forehead to the nape of her neck, and from ear to ear was enclosed in shining sheet gold.

I asked my Guide whatever this head covering was worn for, and his answer was: "Oh, rich ladies wear that kind of thing to show how rich they are, while poor women wear just a band of metal two inches wide, which is sometimes not gold, across their heads to keep their hair tidy." The evening before I saw this golden-pated lady I had arrived at Amsterdam—the Venice of the North—and I thought all the inhabitants had gone crazy. It was the birthday of Queen Wilhelmina, and judging from what I saw, her popularity among her subjects is unquestionable.

All the vehicular traffic in the streets had been suspended and stretched across the streets from pavement to pavement was row after row of dancers, either linking arms or holding hands. I don't think the dancing was very proficient, and almost any kind of a hop or jump seemed to satisfy the dancers.

These rows of dancers were in the majority of cases strangers to one another, and any unattached person who chanced to step on the streets was immediately taken hold of and made to join the revellers, and dance homage to their popular Queen.

On arriving at my hotel at about 9 o'clock in the evening, I expressed a wish to have a Guide to take me out into the town. I went out for about three hours, and found the celebrations still going on at full swing. Needless to say, the cafés and hotels were packed, and I was told by the proprietor of my hotel that no-one in Amsterdam would go to bed that night. Next morning I encountered such a jaded lot of waiters that they hardly had any energy to give me breakfast.

While in Norway I visited some of the very old houses which usually contain one large room with a fireplace at one end of it, serving as kitchen and living room combined. The rule was not to permit any females in the half of the room nearest the fire. They ranked, apparently, as a lower grade human being than the males, and had to remain in the background while the latter enjoyed the warmth and comfort of the fire.

Another example of the low esteem in which women were held I encountered in Belgium. It gave me a shock to see in the towns and cities, light carts each drawn by a woman and a dog.

The woman would be pulling the cart in front and the dog would be behind her under the cart, wearing harness which was attached to the underside of the cart. Literally, the woman was just a beast of burden.

I have already given an example of this attitude to women, when I referred to the Jewish Synagogue in Prague. When returning from my visit up the Hardanger Fiord I stopped at a place called Norheimsund, where there was an exceptionally fine hotel, and where I hired a car to take me to Tregerheid, a station on the Bergen-Oslo railway.

About the middle of this journey the road was blasted for a considerable length out of a granite precipice, and was little wider than the car-track. In common with what I had encountered in many other parts of the Continent, there was no protecting wall or fence of any kind on the outer side, but stones about the size of a baby's head, so that presumably if the driver of the car felt his front wheels bumping the sides of one of these stones, he would know that he was not more than one foot from catastrophe.

About the middle of the length of this dangerous road, it runs at nearly right-angles, and the car had to be driven very slowly round the corner. I observed that when the outer front wheel was just skimming the stones on the outer edge on the left-hand side, the body of the car was almost touching the rock face on the right.

Shortly after passing this corner I looked back and saw that the roadway was actually overhanging the rock below. When I got back to Bergen I met Cook's Agent on the station platform and asked him if there were many accidents on this road on the precipice wall, and he shook his head and said: "Yes, a good many, for very often the wheels come off!" I cannot yet make up my mind after my own experience of that road, whether the Agent was gently pulling my leg or not.

25. INSTRUMENTS OF TORTURE

Now, however, I will tell of some of my gruesome, or near-gruesome experiences. At Ghent in Belgium I was taken to a Cemetery where I saw a number of graves that had been dug to a depth of from five to six feet, in each of which a coffin had been placed resting on the bottom, and with the lid removed. The graves had not been filled up, but at the top of each grave, at ground level, a sheet of plate glass had been fixed so that the relatives and friends of the deceased could look down upon the corpse underneath and see how the worms were getting on with their grisly work.

In the old castle of Nuremburg I saw a collection of the old instruments of torture that were in use in the Middle Ages. Among these was the notorious Iron Maiden. This was a huge figure of cast iron of a maiden. The front could be opened by two hinged doors about six feet high.

The iron in these doors was about two inches thick, and there were a number of spikes pointing inwards, cast into them.

The doors were open and the victim made to stand inside, after which the doors were closed very slowly. Two spikes were opposite the eyes, two others opposite the breasts, and others down the body to the feet. The closing of the doors was controlled by a windlass, and were closed so slowly that it took twenty-four hours before they were completely shut. The effect on the victim inside hardly bears thinking about.

Another instrument of torture consisted of two bars of iron about five inches in length, which were placed across the thumbs of the victim, one above and one underneath, after which the two bars were screwed towards each other. The upper bar had sharp projections, thus ensuring that a good hold would be obtained against any struggle on the part of the victim. As a further refinement, a chain was fastened to the centre of the lower bar and was connected to a peg in the wall in such a manner that the victim's feet were just off the ground, and the whole body was suspended by the trapped thumbs.

A third was a kind of wooden bed or seat on which the man was put to recline with his legs stretched out to a limited extent. The seat was, however, too short for the legs to be stretched out flat, so that the knees were slightly bent upwards. The torturer was provided with a cast iron ball fitted with a handle like a hammer, and with this he would smash first one knee and then the other, until the legs were squashed straight. There was a period of two hours between doing one leg and the other. After that a longer period was allowed, when operations were begun on the shoulders in a similar manner, and so on to other parts of the body until the man was dead.

A fourth, and I think almost even more terrible instrument of torture was the rocking horse. In appcarance it resembled a rather large-sized child's rocking horse, except that the centre-line along the back bone was made rather sharp, and of course of iron. The rockers, instead of being curved in the usual way, had flats in them so as to give a jerky motion when the horse was rocked. The victim was put on the horse, with heavy weights manacled to his ankles. The horse was then rocked as described, and the victim was gradually split in two by the sharp upper edge of the horse. One shudders to think of humans depraved enough to take sadistic satisfaction from such practices, but let us not condemn the people of olden times too sanctimoniously. The nature of the beast still remains, only the methods have changed. It is well to glance at occupied Europe, and ponder on the things that are daily happening there, and to see that such things can never happen again without retribution being exacted.

26. THE PEOPLES OF EUROPE

It might be interesting at this point to give a word or two about the character of people I met in the various countries of Europe. In Norway, they were definitely kindly and well-disposed towards the British, and the same can be said of the Swedes, to perhaps an even greater extent.

During a twelve hours journey from Oslo to Stockholm, my wife and I decided to break our journey at Karlstad, where there was a luxurious and beautifully-situated hotel in the midst of splendid mountain scenery. Strange to say, there was no bus at the station to convey passengers to the hotel, but instead a number of gorgeously-liveried porters was sent by the management to carry passengers' luggage.

I have nothing but good to say of the service we received here, nor of the comfort of our quarters, and the lovely view we had from the windows, which overlooked a wide river used for floating logs from the forests which are a feature of Sweden.

In the morning the hotel hall-porter—a very important personage—cordially shook hands with us, wished us "Goodbye!" and hoped that he would see us again at some future time. I then noticed a very charming lady standing at the entrance to the hotel, and as the porters had gone ahead with our luggage and I was not very sure of the way, I asked the lady if the road I pointed to was the correct way to go, and she replied: "Oh, I am coming with you!" It appeared that she was the hostess of the hotel, and during the journey to the station, and for twenty minutes on the platform she stayed with us. No friend seeing off a guest could have been more amiable, and we were made to feel throughout our short visit not that we were customers but very welcome friends. The lady spoke perfect English, was obviously well educated and cultured, and was not to be wondered at when I learned she was a member of the old Russian aristocracy who fled at the time of the Revolution. This lady had been able to escape with her life, but none of her property, with the kindly intervention of the Lithuanian Ambassador.

At Stockholm we encountered the same kindly sentiments and treatment. We were again treated as almost bosom friends of the hotel proprietor, and one had merely to indicate a desire and any amount of trouble was taken to satisfy it. For example, I only happened to mention that I liked Huntley and Palmer's rusks, and without my knowledge they scoured the City to get them for me.

When visiting the Museum at Stockholm I was deprived of my walking stick, and my sight being defective, and wearing bi-focal glasses, I fell down some stone steps and, besides receiving a severe blow on the head, broke some bones in my right arm. I nevertheless finished my inspection of the Museum before returning to my hotel.

A doctor was telephoned for, and he took me to the hospital to be X-rayed. My arm was then put into Plaster of Paris and held up by a sling. This was a severe handicap for the rest of my tour, but did not prevent me from seeing all that I wished to see. I mention this particularly because of the almost unbelievable kindness of the doctor who fetched me from the hotel in his own car, brought me back again, and the service I received from the hospital together with the Doctor's professional attention for an exceedingly moderate fee confirmed my opinion that in the Northern countries inhabitants are not grabbers.

For a commodious and well-appointed bed-sitting room, large vestibule for hanging up one's clothes, etc., we were charged only twelve shillings a day, and for food 3/6 per day each, making 17/- per day for my wife and myself. The food was absurdly cheap, whole chickens being served at one meal as a portion for one person. These chickens were young and deliciously tender. In fact, the cooking in Sweden was a revelation.

At Copenhagen I also found a courteous and kindly atmosphere. The proprietor of the Tourist's Hotel where we stayed made it a practice during dinner each evening to walk through the dining room with a smiling face and make a courteous bow to each of his guests—a pleasant, friendly custom and characteristic of the Northern countries generally. Moreover, no matter what dish was served, a second portion was always placed on the table beside one's plate, so that if one desired one could help oneself to more of any particular dish. A very different spirit to the mean, grudging spirit I have encountered elsewhere.

In Belgium, I think I may describe the people as intermediate between the kindliness of the Northern countries and the disagreeable qualities I have met with in some others farther South.

In contrast with the liberal way in which food was served at Copenhagen, I may mention that in the carving of a chicken the body was sliced right through bones, meat, etc., thereby doubling the number of portions that could be got out of a chicken.

In France, the treatment met with in hotels generally, as regards the proprietors and managers was just a relationship between an hotel and its customers, with nothing savouring of friendliness. The hotels in Germany were usually very imposing and well-equipped, and provided you could pay with the old Cook's coupons, costing 10/6 per day for room, and all meals, you got good value for money. But I think there were few German hotels that were not out to cheat their customers if they felt they could get away with it.

For instance, if one stays a few days the commencement of the stay will be dated back one day, in the hope that it will not be discovered. Restaurants are associated with hotels and you will be invited to take your dinner in the restaurant, instead of in the hotel dining-room, and a bill for exactly the same kind of meal will be charged at double price, and payable in cash.

Buses will be charged at 2/- for coming, and the same for going to the station, where the hotel may be just across the road, and one has not used a bus at all. I have been told that the reception clerk, when booking rooms for a customer, has a secret book in which he or she records what I may call the "cheatability" of the customer. I definitely do not regard German hotels or hotel proprietors and managers as being either friendly or considerate towards their customers.

In Switzerland, I cannot say that I found any tendency to cheat customers that patronised their great hotels, nor on the whole have I found them an unfriendly people. In Italy the hotels were fairly comfortable, and on the whole one got a square deal, but nevertheless the standard of honesty of the whole country is low.

I would like to deal now with another aspect of the treatment and service one receives in the different countries, and to a great extent they follow the same lines as the general character of the people.

During my journey from Odda to Stallheim in Norway, I came across a museum containing examples of Norwegian clothing, implements, arms, and curiosities extending over a number of centuries. The door was locked but there was a notice outside stating the key was at a farmhouse a little distance away. A young woman from this farmhouse came with us to unlock the door, and I spent two interesting hours in the museum, during which time the young woman waited for me to come out, so that she could relock the door when I left.

I offered her the equivalent of half-a-crown, which she refused to accept. I then put a number of coins of different denominations on my outstretched hand and asked her to pick from them whatever she chose. She took the equivalent of 2d for her two hours wait and positively refused to take any more when I pressed her to do so.

At the hotel at Odda, at which I stayed both on my journey to and back from Stallheim, I received a good deal of valuable assistance and information from the hotel boots. I had not tipped him on my way to Stallheim, and therefore I felt that a more generous tip was due on the return journey. I knew that large tips were not expected, so I put the equivalent of 4/- out on my hand, and said: "Take what you like. The whole of it if you choose." He picked up three coins and replied: "That is quite enough for what I have done for you, sir." It amounted to about 2/- of our money. That could not have occurred in either France or Germany.

I will here make a digression to relate a remarkable incident that occurred in connection with this man. I remarked that his English was extremely good, and he replied: "Oh, I go to England every year for my holidays. I have just returned from one now." "Indeed!" I answered. "And where did you spend it?" "At Aberglaslyn Pass," was the answer.

"Well," I said, "that is only four miles from my home!" Odda is at the extreme end of the Hardanger Fiord in the heart of Norway, and I had met there this man, who in the ordinary course of things might not have been expected to travel far from home, yet who had spent his holidays within a short distance of my own home village.

I have often in Norwegian hotels found difficulty in getting hotel employees to accept reward for their pleasant and genuinely friendly service. There was never any standing about to waylay guests in the expectation of receiving tips, a feature that is so annoying in one's general hotel experience in most countries. What I have described in detail for Norway was true of hotel servants, railway porters, and taxi drivers in Sweden and Denmark, but these classes of employees are intermediate in Belgium, both as to their character for service and their greediness for tips.

In Berlin I had a most extraordinary experience at the Hotel Prince Albrecht. On leaving, when I came to the vestibule of the hotel, I found nine hotel servants in a line on my left, with the Hotel Manager standing in front of them like a drill sergeant, and I had to pass between the line of servants and the Manager. About half of these servants I had never seen before, and I was not aware that they had ever done me any service, but I had to place a tip in each outstretched hand, and after I had finished the ninth, the Manager approached me.

"Have you seen the chambermaid?" he asked, and I was able to inform him that I had already provided a tip for her in the bedroom. I tipped, I think, every hotel servant that had as much as looked at me. Even so, the bus conveying me to the station had gone but thirty yards when I saw two men running like hares after the bus. They caught up with it and jumped on the step at the end, and the first panted: "If you please, sir, I am the under-boots!" The other fellow, not to be left out, wheezed: "Please, sir, I am the man that cleans the knives!" They had actually run after the bus to demand their tips. Contrast this with the behaviour of hotel servants in the Scandinavian countries.

Now for France! I will say at once that railways porters, taxi drivers, and waiters or other hotel servants are detestable. It matters not how liberal your tip, they will put it on the palm of their hand and almost push it into your face, and try to argue with you that it is totally inadequate. In my opinion they are nearly all of them twisters of the worst type.

To begin with, hotel servants demand 20% of the bill in tips, whereas in no other country is more than 10% expected. On one occasion I returned to the Gare el Est a little after midnight. I had thoughtlessly parted with my small change to a fellow-traveller who had asked me to change a 20-franc note for him, leaving me with no small money for porters or taxi-drivers.

I mentioned this to him, and he said: "Oh, we will put all our luggage together, and I will tip the porter." When the train drew up at the platform, instead of there being porters all along the train, as one would find in England, there was not a single porter in sight. We took the luggage out of the carriage ourselves, and leaving my wife and fellow-passenger to guard it, I went in search of a porter, found one, and got him to bring the truck along for our luggage.

When we arrived at the station entrance, my fellow-passenger very foolishly gave the porter his tip, and then asked him to find a taxi for us. The porter merely walked across the road and came back again. "You cannot get a taxi tonight," he said, and marched back into the station leaving us in the lurch.

My companion then said he would go and find one himself, if I would stay with the luggage.

There was a fog on at the time, and my fellow-passenger soon returned with failure written on his face. While he had been away the porter had returned to where I was standing and told me that my fellow-passenger had not given him a big enough tip, and that I must give him some more. This I had refused point blank, at which he threatened to take my fellow-passenger's luggage back to the platform whence he had brought them. Indeed, he had actually started off with two suit-cases before I managed to grip him by the shoulder, and give him a shake, and a murderous expression, which I think somewhat frightened him. Anyway, I managed to safeguard the luggage, so that when my friend returned, I was able to leave him in charge and go out to try and find a taxi myself. This I succeeded, but could not find a second, so we arranged that the three of us should share the one taxi, the driver taking my friend to his hotel and then continuing the journey to ours, with my wife and myself.

We dropped our fellow-passenger at his hotel, and had only travelled a short distance when the taxi stopped. Suddenly, in almost complete darkness, the door of the taxi was flung violently open, and hands groped in and pulled all our luggage out on to the pavement. We had eight rather heavy pieces. I jumped out and found these already in the possession of four or five villainous looking men. As quick as lightning I ran to the piled up luggage, and brandished my stick, my wife standing by me. Turning round I saw dimly in the distance the name of our hotel illuminated by electric signs. I ordered the taxi driver to complete his journey, first helping me to get the luggage in again, but this he refused to do, and I have no doubt that he was in league with the other rascals. Anyway, whether it was my angry bearing or my minatory stick that was responsible, I don't know, but the band suddenly decided to give me best, and scooted, taxi and all.

I had not long before undergone an operation, and had been told I must not lift heavy weights, but I had to put a suit-case under each arm, and hold two other pieces in my hands.

My wife did the same, so that between us we toted the eight pieces in one journey to our hotel.

Bearing in mind that it was now about 12.45am, I think this experience of railway porters and taxi drivers in Paris justifies what I said in the beginning about them.

Restaurant cars ran regularly between Calais and Paris, and the meals consisted of several courses. Invariably, before one had eaten one-third to one-half of the course the French waiters would snatch the plate away, and one would find oneself at the end of the meal still hungry, because one had not been permitted to finish eating the not-too-plentiful platefuls supplied which had been ordered and paid for. Probably the motive was to rush as many diners through the cars as they could in a given time, and thereby increase the takings and their tips.

If one asked for plain water to drink, expecting to get it free of charge, as in England, a bottle containing plain water would be brought and never put down on the bill for less than 1s 3d of English money. I resented the snatching away of my food, and always looked out for the waiter's attempt to take it so that I could hold on to my plate and finish eating what I had paid for. There was generally an expression almost akin to murder in the waiter's eye on these occasions, but I risked the consequences. I can hardly express the feeling of satisfaction I used to have when I arrived back in England with its polite, cheerful and helpful porters and waiters, and to meet again people whom you did not have to suspect of ulterior motives in their attitude and behaviour towards you.

I was well acquainted with Lady Arbuckle, who was a charming French lady, and the wife of Sir William Arbuckle, the Agent-General of Natal Colony in this country. At one time, after visiting her native country, she came to stay at my house, and told me about the Bolshevist views which she had met with while in France. A lady had said to her: "It is not right that you should have interest on your money," to which Lady Arbuckle had replied: "Then we would not lend it to you. Why should we?" "But what if we were to kill you, and then take all your wealth, my friend!" Such was the calm rejoinder, and from what I have experienced in France, I believe that this is a true reflection of much that is thought by people in that country who should know better. It expresses a low sense of justice or morality.

During my tours in Europe I was tremendously impressed by the advantages enjoyed by the different countries in comparison with our own by their adoption of the decimal coinage, and the metric system of weights and measures.

Take our stupid and out-of-date monetary system. If we wish to add various sums of money together, we must first add the farthings and divide by four to get the pence, and the remaining farthings if any. Then we add the pence and divide by twelve to get the shillings, and remaining pence if any. Then we must add the shillings and divide by twenty to get the pounds and the remainder of shillings. Finally we add the pounds.

Note that there are three divisors—a four, a twelve and a twenty, together with three subtractions in order to get the correct remainders. In the decimal system which prevails on the continent, a simple addition sum is all that is necessary. There are no divisors, and no subtraction sums to get remainders which we have to enter as pounds, shillings and pence, and often enough even farthings. As it is, our banks find it better to confine their business to the nearest penny, it is not worth clerks' time to bother with halfpennies and farthings.

To show how little change would be necessary to convert our idiotic monetary system into a logical and convenient one, all that would be required would be to slightly depreciate the value of the farthing, and divide our pound into 1000 instead of 960 parts. The florin would thus contain 100 farthings, the equivalent of our present shilling would be 50 farthings, and would be stamped 50. The equivalent of our present 6d would be 25 farthings, the "near" equivalent of our 3d bit would be 10 farthings, or the tenth of a florin, and would pretty well correspond in value to the American nickel. Smaller denominations could be arranged as would be most convenient.

Not only would this system be easy for those who have a limited education, but it would enable the slide rule to be used in business offices to a very great extent, and no one who has not had experience of this instrument can conceive the enormous savings in time that is obtained by its use in making long and complicated calculations. Supplemented by the metric system of weights and measures, which is well adapted for the slide rule, I do not hesitate to say that 40% of the time now spent in our counting houses could be saved.

But to return to the peoples of Europe! Regarding Italian hotel servants, porters, etc., I have not much to say either good or bad. There is, however, one little country that I have not yet mentioned, and that is Czechoslovakia. I once stayed at the Schwarz Ross Hotel in Prague for a week, and a friend of mine, a resident of the City, asked me if I knew anything of the rules of hotel tipping, and I said that I did not.

"I will save you money," he offered, "if you will allow me to pay your hotel bills for you." I accepted his kind offer, but was later surprised to find what a large amount relative to the total bill, had been given in tips. My friend then explained to me that it was the rule for any hotel meals, to give three tips—one to the head waiter, one to the waiter who served at your table, and one to the boy who washed the dishes.

The tip to the head waiter must be placed in his hand. The tip to the waiter must also be placed in his hand. Money for the boy who washed the dishes was left on the table.

Before concluding my continental experiences, I would like to comment on the extraordinary way in which Americans "see Europe." I travelled with my wife and brother from Frankfurt to Berlin, and a fellow passenger named Mr Wolf, with whom we were on friendly terms was with us throughout the journey, and stayed at the same hotel in Berlin. Next morning, after breakfast, we went to Cook's Office to see what arrangements we could make for seeing Berlin.

Mr Wolf, as soon as he entered the doorway, addressed a stolid German standing behind the counter. "Mornin', Colonel! I want you to tell me how to see Berlin in a day!" "You cannot see Berlin in a day," was the stolid answer. "You cannot see Berlin in less than a week." Mr Wolf answered this airily; "My good fellow, I've got to see all Yurrup in a week!" The upshot was that we had to hire a taxi for the day and dash around Berlin to the various places of interest. That, at any rate, was the only practicable solution, so we went to the street and hailed a passing Jehu. He could not speak a word of English, and we were without German, so it was difficult to communicate our wishes to him.

However, we got in, and before long found ourselves being driven through an apparently endless forest of trees. Mr Wolf became restive, and said that he hadn't come to Yurrup to see god-dam trees and grass, but nothing we could do would make that thick-headed Teuton understand our lack of interest in forestry.

At last my companions took to poking the taxi-driver in the ribs with their sticks, until at last he looked appealingly at me for succour. It took some time to make him understand that we wished to get back to Berlin from these woodland environs, and glad we were to be able finally to dismiss him, an hour and a half of precious time wasted.

After that we walked about the streets at random, returned to the hotel, had lunch, and—will it be believed?—our Mr Wolf started for Vienna by a train that left Berlin at 1.0 pm Mr Wolf had "seen" Berlin.

Another time I was at Dresden Picture Gallery, where, at the entrance the Great Hall is filled with pictures by Rubens. A party of about six American women stalked into the Hall, headed by a masterful-looking female, with a notebook in her hand. She walked up to the first picture, gazed at her note-book and ticked it with a pencil, and with but a brief glance at the immortal masterpiece, moved on to the next. Again the formidable pencil stabbed the listed page, and so it came about that Rubens became properly "ticked off" in the end, and the grim woman stalked out, followed by her meek disciples, having "done" the Great Hall in less than five minutes. I suppose when they got back to the USA they would look for the ticks in the note-book, and verify that they had "seen" all those lovely pictures in the Dresden Gallery. *Vita brevis, ars longa* (Life is short, Art is long).

It appears to be a common practice in America to book tours with the American Express Co, and I have known middle-aged American ladies, the itinerary given them, simply taking the train journeys, going to their selected hotels, staying there until it was time to start on the next stage of their journey by rail, but seeing nothing of what was interesting in the places they passed through. Yet I assume that these good ladies are convinced that they have "done" Europe thoroughly. One thing is sure, somebody was being "done." While we were at Copenhagen a party of Americans arrived at our hotel, saw Copenhagen in three hours, and then started off for another place, much to the disgust of the patriotic hall porter who vowed that Americans never really saw the worth-while places of interest in the Capitals they visited. I think, when such people get back to their native land, the jumble of impressions they have collected must leave them in a very confused state of mind.

It reminds me of the story of the American lady who had done Europe, and who was therefore supposed to be an experienced traveller. She was consulted by a friend, who also proposed to do Europe, and this was the advice she gave: "Now, when you are in England, there are two things you must be sure not to miss. One is the Bank of England, and the other is the Church of England!" Coincidences of a remarkable nature have occurred to me during my travels, thus confirming the old truism that the "world is a very small place." Mr Wylie, a Director of the Port Meirion Hotel, once took a small cottage at Croesor, called "Gattws", and which was formerly the lodge of the Park House. He invited me to take tea with him at the cottage, and there I met his friend Mr Peter Luling.

Three weeks later, at Cortena Arpezzio, in the very heart of the Dolomites, I met these two gentlemen on the street. This was at a spot about a thousand miles or more from Croesor but this does not exhaust the strangeness of the coincidence.

The two friends had left England unknown to one another, Mr Luling to Munich, and Mr Wylie to Innsbruck. A telephone ring at his hotel bedroom was the first intimation that Mr Luling was anywhere near, his friend having discovered that Mr Wylie was at Innsbruck. So each then varied their proposed itineraries, and agreed on Cortena as a meeting place. They had only been there about an hour before they saw me, and the three of us who had had tea together at Croesor a short while previously, had tea together again at Cortena, each of us having arrived at our meeting place by widely divergent routes.

In March of that same year, I received a letter from my cousin, Mrs Minnie Masters, saying that Mr James Masters—Uncle Jim—would be coming to England shortly, and proposed to call on me. I heard nothing more until the beginning of the following August, when I received a letter from him from an address in Somersetshire, asking if it would be convenient for him to come and stay with me.

I had then booked the tour through Venice to the Dolomites, etc., and was due to start on the following day. I wrote suggesting that he should join me in this tour, but he did not do so, as the notice was too short. He said, however, that he would arrange to spend some days with me in London, after my return. We were then out of touch for several weeks, during our separate travels.

Prior to returning to England I decided to stay a few days in Paris, and on the day following my arrival in that City, I went to the Palace of Versailles, which is some considerable distance from Paris. I spent the morning going through the picture galleries, and then went to a restaurant in the town of Versailles for luncheon.

Luncheon ended, I walked leisurely out of the restaurant, and going along he street became conscious of two other people who were walking beside me, talking together. It was my Uncle Jim and his wife. I had no idea that they had left England, and he explained to me afterwards, that he was bitterly disappointed that he had been unable to accompany me on my extended tour, and as a compromise he had decided to visit Paris. It was a mere whim which had decided them to see Versailles on this particular day Had I loitered a little more over my luncheon, or been just a little more brisk, I would probably have missed this chance meeting altogether.

27. MANUFACTURING PROCESSES

In the course of my career I have had the opportunity of inspecting many important works, and witnessing processes of a varied character. In Dolgarrog I saw in detail how aluminium was manufactured from bauxite. This mineral was imported from France, and cost, I was told, about 15/- a ton. Now it is interesting to note that slate contains about 22% of pure aluminium, and that the huge rubbish heaps which are a feature of Blaenau Festiniog and other quarry districts of North Wales, contain aluminium to the value of many millions of pounds, But the process of extracting it on an economic basis has yet to be discovered.

I saw aluminium being manufactured in all its processes at Dolgarrog. There is little doubt that with the big developments in aircraft and consequent demand for light metal the manufacture of aluminium will assume huge proportions in the world.

There is an interesting and comparatively new alloy of aluminium called electron. It contains a large percentage—approximately 50%—of magnesium, and its specific gravity does not much exceed one-half of that of aluminium. I first saw it at an exhibition at Bergen, Norway. The exhibit was in charge of a woman, and when I asked her what the metal was made of she said: "The foam of the sea." I thought the "foam of the sea" was a rather peculiar source from which to derive any industrial metal, but no amount of questioning would get that woman to admit that it was made of anything more substantial. The chief objection to it is its inflammability, even the friction of machining processes may set it on fire.

I was always very curious to know how very long paper-insulated cables were enclosed in a lead pipe, so as to protect it from damp, and this without transverse joints.

At the works of Messrs WWT Glover & Co, Trafford Park, Manchester, I saw this being done. The paper insulated cable was being drawn through a machine and simultaneously a mass of lead slightly softened by heat was forced through a die, which flattened it out to the thickness required in the lead pipe, and then wrapped it round the paper insulated cable. The edges, not having become oxidised, cohered, and formed a water-tight and permanent joint, running the whole length of the cable.

I may mention as a matter of curiosity that at the works of Thomas Hancock & Sons, Sheffield, I saw files being cut by hand. This was done by taking an annealed bar of smooth cast steel and laying it upon a leaden anvil. The workman who was cutting the teeth in the file had a chisel a little wider than the bar of steel, and a mallet or hammer with which he struck the chisel heavily enough for it to penetrate the steel bar.

After each tooth was cut, the chisel was moved by hand to the position for cutting the next tooth, and the remarkable thing about the process was that even with the finest files no measurement was taken to enable the teeth to be cut at equal distances from one another. In fact, the chisel seemed to find its correct position on a rebound after a blow, and they were cut with such regularity that to the naked eye they seemed to be absolutely equidistant. There is no doubt that these file cutters have developed a wonderful sense of touch, and instinctively moved the right amount each time, and also maintain a regularity of angle of the teeth across the face of the file.

I noticed that notwithstanding the extreme accuracy with which the files were being cut, the workman was not even looking at what he was doing.

I saw ball-bearings being manufactured at the works of Hoffman & Co, and as indicating the extreme accuracy obtainable by modern methods of production and high degrees of training in the workmen engaged in producing the articles it may be mentioned that the balls and ball races are accurate to less than one ten-thousandth of an inch. When it is considered that one thousandth of an inch is represented by a very thin sheet of tissue paper, and that one ten thousandth is but one-tenth of that thickness, the degree of accuracy can be appreciated.

In machine tools intended for ordinary machining, remarkable accuracy is obtained. At the works of Dean, Smith & Grace, who are makers of high-grade lathes, I saw a long lathe-bed upon which lay a steel straight-edge, and at intervals of six inches little slips of tissue paper, one-thousandth of an inch thick, were placed between the straight-edge and the lathe bed. I was invited to pull the end of each of these tissue paper slips, but found them to be tight along the whole length of the bed due to the weight of the steel straight-edge.

A deviation of even a small fraction of that thousandth of an inch along the whole length of the bed would have made one or more of those tissue papers slack. Of course, wonderful accuracy can be obtained by grinding processes, and it is astonishing what complicated forms grinding can be successfully applied to.

At the works of Thomas Firth & Sons, Sheffield, I saw huge bars of steel comparable to large baulks of timber heated to a red heat, and being passed between rollers like a monster mangle, and flattened out several inches at a time, almost as if they were dough—no less than 6000 horse power being applied to turn the rollers. But even this large horse power is not sufficient to roll the heavy section required in shipbuilding and other structural work, and great manufacturing country as we are, we had to buy our heavy sections of steel from the USA, where they have the enterprise to put in large and powerful machinery where such is required.

This stigma has recently been removed by the English firm of John Summers & Co, working in conjunction with an American firm, and the largest sections of steel required for structural work are now being produced in this country.

Very advanced manufacturing machinery is to be found in such works as Alfred Herbert & Co, of Coventry, particularly in their capstan lathes, and their advanced designs in milling machines. I have seen extremely rapid work performed at their factory, and the capstan lathes are typical and the very essence of mass production, which, notwithstanding the sneers of the unprogressive antediluvians, is the only method of manufacturing worth considering where large quantities of any article are required. I have also seen fine examples of capstan lathes, profiling machines, and other modern tools at the works of Wards of Birmingham.

It is a far cry from the engineering works and processes I have described, to cotton, but I cannot conclude my experiences of works and manufacturing processes without referring briefly to a visit which I paid, and which I was specially privileged to make, to the cotton mills of Horrocks, Crewdson & Co, of Preston. It was impressive to see the means taken to weave the wonderful patterns in which modern fabrics are made.

I think I can also claim to have been brought into contact with artificial silk at its very inception. I was present at interviews at which the pioneers of the industry were showing neckties and other articles of clothing in which the artificial silk had been superimposed upon a cotton base.

I remember the arrangements for the first Company floated and the tremendous demand for the shares. The would-be shareholders standing in a long queue to put in their applications, yet notwithstanding the intrinsically sound character of the invention this Company was not a financial success.

28. QUARRY CASHFLOW PROBLEMS AND ATTEMPTS TO SOLVE THEM

I have already mentioned that over a fairly long period of years I was shipping Best Old Vein slates to the continent, and especially to Germany, and my relations with my German customer were of a very cordial character.

The shipping of slates to Germany could not be carried out during the winter, but it was usual to book orders for the spring trade in January or even in the December of the previous year. In the early part of 1914, these orders did not materialise as usual, notwithstanding the good relations that prevailed between my German customer and myself, nor were any orders for cargoes received during that year at all.

I am not quite certain as to the cause of this cessation of orders. It may even have been due to a foreknowledge of the coming war, and the payments might in consequence become long deferred. Up to 70% of our total output was Old Vein slate suitable for the German market, and the holding of this substantial proportion of our output from the end of the shipping season of 1913 to July of 1914 meant a heavy accumulation of stock, and a certain amount of financial stringency.

The consequence was that the Debenture Interest due on the 1st July, 1914, was not paid on the due date, and quite at variance with the extremely accommodating attitude of our Bankers, they became much less accommodating, which was a matter of considerable surprise to me at the time, but I now think it was foreknowledge which was probably common to the banking fraternity of the imminence of war, and the consequent conservation of their liquid assets, which was the cause of their change of attitude.

The non-payment of the Debenture Interest on the 1st July, 1914, rendered the immediate repayment of Capital and Interest a legal liability, and one of the smallest of the Debenture Holders started a Debenture Holders' action against the company. The difficulty could certainly have been got over had normal conditions prevailed, but with the general upset due to the outbreak of the Great War it was considered advisable, in order to conserve the assets, to appoint a Receiver, and I was appointed Receiver under the High Court accordingly.

As during the Great War all issues of new capital had been prohibited, there was a scramble for all the capital available in the country, following the conclusion of peace, and unless returns up to even 10% on sound Preference could be shown as prospectively available, new money could no be got. Slate quarrying at the time was not a popular form of investment, and the result was that I was unable to bring my Receivership to a conclusion until 1926.

In 1924 two men from the North of England (whom I will call Mr A and Mr B) came to see me, to ascertain whether I could arrange to supply them with 5000 electrical horse-power from Aberartro at Llanbedr, where I was carrying on a hydro electric scheme. Being unable to do this, I told them of the scheme I had for the development of hydro-electric power on the Park and Croesor properties, but I pointed out that I did not think this could be disassociated from the sale of the properties as a whole, but notwithstanding this the proposition seemed to be attractive to them, and negotiations with that end in view were begun.

At that time the cotton industry was in a parlous condition, most of the cotton mills being heavily in debt to their bankers. I was informed that one of the men with whom I was negotiating was a director in twelve different cotton companies, and in common with other directors a guarantor of the overdraft. I was told that one of these companies owed the Bank £150,000, and the Bank exercising their rights under the joint and several guarantees, chose to pick out one of the directors as their target, and this happened to be one of the men with whom I was negotiating for the sale of the Park and Croesor properties.

This compelled him to file his petition, and he left the country for Jamaica. This loss of one of the two men with whom I had been in contact caused a hiatus in the progress of negotiation, but it was eventually arranged on a basis of deferred payments that the other partner should become the purchaser, the position being safeguarded by the condition that I was to remain in absolute control of everything until the property had been finally paid for.

The scheme this man had in mind was of a very grandiose character. He informed me that a company owning paper mills were on the look out for a site for a new mill, and the scheme was not only to develop all the considerable electrical horse power available by utilising the fall down to the foot of the Croesor railway-incline, but also to run a pipe-line from there to Portmadoc, to carry fresh water required in the manufacture of paper. The mill was to be built on the ballast bank at Portmadoc, and a bridge was to be built connecting the South Snowdon Wharf with the ballast bank. The material for the production of the paper was esparto grass, to be brought from Spain in ships capable of carrying thousands of tons.

Hitherto, the vessels that had come into Portmadoc harbour had not had a carrying capacity of more than about 400 tons, but it was proposed to dredge a channel deep enough to admit the larger vessels, and bring them directly to the mill on the ballast bank. I was told that this company was prepared to spend a million pounds on the scheme. Some whispers of this got about Portmadoc, and vastly excited some of the inhabitants, who thought that the prosperity of the port and town would be thereby restored. Whatever foundation there was for this assumption, it did not materialise sufficiently in cash to enable the purchaser to carry out his obligations.

Notwithstanding the arrangement for easy payments, the purchaser found it necessary to assign his interest under the agreement for sale to another man, whom I will call Mr C, and this man undertook to complete the purchase before Christmas, 1929.

Fresh agreements were prepared accordingly, but he never signed them. The reason he gave to me later on was that during November, 1929, I had had to undergo a serious operation, which ended fatally in the majority of cases, and another was that a solicitor with whom he had arranged to bring in a substantial amount of capital for the development of the properties, had had a paralytic stroke, which rendered him quite helpless and incapable of transacting business.

As a result of this the capital he was to have introduced was not forthcoming. About six months later, after I had made a fair recovery after my operation, I had an interview with Mr C, and the question of resuscitating the agreement to complete the purchase of the properties was discussed and finally agreed to, and the date was named on which the balance of the purchase money was to have been paid.

It so happened that Mr C was engaged in large financial transactions with the notorious Clarence Hatry, and Hatry's failure occurred in the very week that had been fixed for the completion of the purchase of the Park and Croesor properties. Mr C told me himself that his loss due to the Hatry failure exceeded £200,000, and it is almost needless to say that this caused him to default in the payment of the balance of the purchase money due on the Park and Croesor properties.

I then approached the Oakeley Company on Mr C's behalf with a view to their purchasing the Park and Croesor properties, not only for the purpose of acquiring its large slate reserves but also to utilising the hydro-electric power on the property, and taking it to the Oakeley Quarries, and there using it as the source of their electric power.

The Oakeley Company while much interested in the slate and electric power part of the scheme, did not want to encumber themselves with 2,600 acres of agricultural land. There is no doubt the scheme was an attractive one, and advantageous to them, and the objection arising from the fact that the properties included so much agricultural land could have been got over, but here again Fate intervened, and the financial crisis of 1929-30—when the Coalition Government rendered it necessary to conserve capital very carefully and made it inadvisable to embark on any new enterprises such as the purchase of the Park and Croesor properties—would have involved too much risk.

Eventually a cancellation of the contract with Mr C was arranged on terms, and it became necessary for me to find another purchaser. I entered into negotiation with two gentlemen, one of whom was a solicitor, and terms were definitely agreed verbally, but the Solicitor said he would prepare the offer in writing. When the written offer was received, it was accompanied by a letter saying that the offer was conditional upon the County Council making a road from Tanlan to Croesor.

Obviously I could not bind the County Council, and declined to accept the condition. The other partner, however, with whom I was in direct contact agreed to withdraw it, and the formal contract was drawn up and sent to the solicitor partner, who returned it unsigned. Thus another promising negotiation came to naught.

29. PROBLEMS OF ACCESS TO CROESOR

I would like to mention at this point that this was the fifth or sixth time that substantial capital, which would have been available for the development of the Park and Croesor properties was withdrawn at the last moment, owing to lack of proper access.

In one case, a man with £100,000 at the bank, awaiting investment, was prepared to purchase and develop the properties on an extensive scale. All preliminary investigation of accounts by a Chartered Accountant had taken place, but when he visited the properties personally, its inaccessibility condemned it, and he wrote to me, saying: "You have the slate, you have the electric power and every condition of success except one—accessibility, and how you ever get one to go to it, I don't know." And for this reason alone he turned down the proposition.

Considering there was a group of Quarries in the neighbourhood of Croesor—Croesor, Rhosydd, Pant-mawr, Cefnbraich and Park Quarries—it is scarcely believable that no publicly owned road existed which connected them with the outer world.

Before the Traeth Mawr was reclaimed from the sea by the building of the Embankment at Portmadoc, and before there were any Quarry developments, a narrow parish road generally about the width of a cart track had been made from Croesor over the ridge of a hill near Cefn Deriniog, and thence to near the old church of Llanfrothen, and thence along the north side of the hill which lies north-east of Penrhyndeudraeth to Penrhyndeudraeth itself.

Of course, with the reclamation of the Traeth Mawr such a roundabout route was an anachronism, and parts of it disappeared altogether. There was therefore no access on any publicly owned road along which a vehicle or even a pedestrian could reach Croesor from the outside world.

The owners of the Croesor Quarry for their own convenience, and at their own expense, made the wide road which extends from Rhiw Ty Mawr, Penrhyndeudraeth to Garreg but the difficulty was to link up this road with the old parish road previously described. The owners of the Brondanw Mansion had a private carriage drive leading from this lower road to their house, and beyond this they owned a narrow lane which ran from the entrance to the Brondanw Mansion to a point on the old parish road about 1½ miles from Croesor.

This section of the old road was known as the Coed Ty road, and with extreme courtesy they permitted the use of these roads by the public, but subject to withdrawal at any time. They absolutely refused to permit these roads to be made public roads, or to be tampered with, and this meant that the narrow lane above Brondanw which ran between walls about six feet high, was a bottleneck of a very dangerous character. It was impossible to take a motor car along it without risk of damage, unless the driver was skilful.

If a pedestrian was met in this lane he or she had to climb to the top of a six foot wall, or run back to a wider place much farther down, in order that the car could pass. Added to this, the old Coed Ty road fell into such a state of disrepair that no vehicle could travel along it. I was informed that for more than twenty years not a penny had been spent on it. The walls at its side had fallen into the road and almost completely blocked it, and for considerable parts of its length it had become covered with brambles to such an extent that it could only be used as a path.

The workmen at Rhosydd Quarry living in the lower parts of Llanfrothen made an appeal to the Llanfrothen Parish Council to cut away the brambles so that they could walk along it, but nothing was ever done. This description does not depend on hearsay, but upon my own personal observation and knowledge of the conditions, I living at Croesor at the time.

Obviously such a state of affairs could not have prevailed for any great length of time unless some other means of getting to and from Croesor was available. This other means consisted of a privately owned road leading from the main road by Tanlan, through a gorge called Ceunant to the old Park Mansion or Farm House, and here again the public were entirely dependent upon the goodwill of a private landowner to enable the villagers to get to and from Croesor to the outside world.

When the old Park Mansion was in its glory, the road through the Ceunant, which was 24ft wide, was a fine carriage drive, and though only part of this was in use during the time I knew it, the foundations of the old 24ft roadway were there, and the road could be restored to its original width at small cost.

The fact that this road joined the main road near Tanlan, which was the nearest point to the main road that came to Croesor, made it equally suitable for traffic coming from Beddgelert direction, as it was for traffic coming from Penrhyndeudraeth or Portmadoc direction, and this was obviously the starting point from which a public road to Croesor should have been made.

In the absence of a road leading from this point traffic from the Beddgelert direction, would have to travel nearly a mile to Garreg in order to enter the private drive of Brondanw Mansion, and then double back to a point above Tanlan, and after a journey of nearly two miles it would be only 150 yards on a plane from where it was before, and then of course it would have to traverse the old Coed Ty road, provided the latter was rendered safe and usable.

I have known funerals where it was necessary to take the hearse down the Park or Ceunant road, and then up past Brondanw and a narrow lane to the Ramoth burying grounds—a journey of about three miles instead of one, if the Coed Ty road had been usable.

It is scarcely credible that public authorities, whose duty it was to provide road facilities for every village and town in the County, should have been so utterly neglectful of their obligations as to allow this intolerable state of affairs to continue. Yet, nevertheless, owing to personal animosities and misconceptions this kind of thing practically prevails today, and the development of a Welsh village possessing potentialities that might have made it almost a second Blaenau Festiniog has been prevented.

Notwithstanding the substantial Government help that is now available for the carrying out of such schemes, matters remain in the same negative state of inactivity. I hope to return to this.

I will add just one more instance of where capital was alienated by the lack of proper access.

A wealthy capitalist from the North of England came down by car to see the properties, with a view to purchasing and developing them. He travelled via Beddgelert, thence to Garreg, and up the Brondanw carriage drive, but when he came to the narrow lane he said: "If it is up a place like that, I am going no farther." He turned back his car, and returned home without visiting the properties, although he had already completed more than 99% of his journey.

The withdrawal of the solicitor partner from the deal which I have already referred to upset the negotiations with the two partners, as the remaining partner felt that the obligation to find the purchase money and also capital for development was more than he was prepared to undertake. However, he told me that if I could see the way to find the capital for development, he could find the purchase money.

He also made a stipulation that I was to continue my association with the undertaking, and help him with guidance and advice during the early stages of development. This I agreed to do, and I got in touch with a very responsible financial firm in London, who entertained the proposal very favourably. The amount I wanted to raise for development was £100,000, and I met at their offices gentlemen who were prepared to find this money.

I had brought the negotiation up to a point at which an interview between my Principal and the prospective investors was desirable, and an appointment for this purpose was made at the Office of the financial firm referred to. Members of this firm, the intending investors, and I, turned up for the meeting, but my Principal, whom of course we had expected also to be present, sent a telegram saying that he had resold the property.

The new purchasers bought through a nominee, and for some time I was unaware of their identity. They asked me to keep things going on until permanent arrangements might be made, which I agreed to do, but being in a very low state of health and desirous of retiring, this arrangement came to an end after about six months. The new purchasers, whose identity I eventually discovered, were one of those powerful financial houses in Great Britain, but ruled out of their investments mining and foreign investments, and their idea was to develop the Park and Croesor estates as agricultural estates.

In comparison with the mineral the agricultural value was exceedingly small, and they eventually sold it to local gentlemen who were much more likely to develop all its resources.

30. HYDRO-ELECTRICITY

Early in my association with the Park and Croesor properties I realised their potentialities for the production of hydro-electric power, and had worked out schemes having that end in view. It was, however, between the years 1920 and 1930 that I fully developed these schemes in all their detail.

The first scheme was to (a) raise the height of the dam at the Cwm Foel reservoir so as to considerably augment the storage capacity of the Cwm Foel valley, (b) utilise the remainder of the catchment area not already utilised, which roughly comprises land lying to the north of the railway leading to the Rhosydd Quarry, and (c) the utilisation of the catchment area on the south or Moelwyn side of the aforesaid railway, by means of a small reservoir near the Rhosydd railway, and connecting it by a pipe-line with the hydro-electric station at Blaencwm.

Arrangements were also included in the scheme for utilising the ample storage capacity in the Cwm Foel reservoir for conserving the surplus water on the Moelwyn side during spates.

The second scheme was of a more ambitious nature, and involved the acquisition of the little additional land beyond that already comprised in the Park and Croesor properties. There is near Garthllwynog Farmhouse a rock about 60 ft. in height, which juts out at right angles to the longitudinal line of the Croesor Valley, and two-thirds closes it at its lower end. The scheme comprised a dam constructed to close the remaining third, and the road and dam would give a reservoir no less than one and a quarter miles in length, and the surface of the water when the reservoir was full would be not much short of a square mile.

A pipe-line following approximately the course of the Croesor railway would be laid from this Croesor Valley reservoir to the foot of the Lower Park incline, giving a head of nearly 600 ft. when the reservoir was full.

Water from the Maescwm Catchment Area comprising the west side of Moelwyn Mawr, and the south side of Cefnbraich, would also be brought to the Croesor Valley reservoir, thereby augmenting the supply. The carrying out of schemes of this kind would naturally depend on how the current generated was used, and what price could be obtained for it, but as the North Wales Power Co.'s transmission line passes through the Park property only about three-quarters of a mile from the foot of the Lower Park incline, where the main generating station would be located, the basis of a promising negotiation with the North Wales Power Co. to purchase the current generated, appeared to be present.

Failing this, however, there were other means of utilising electric current in the manufacture of commercial commodities such as aluminium or carbide of calcium. The total amount of electricity that would be rendered available by both schemes exceeded five million units per annum.

There was, however, an exceedingly active negotiation in progress with the Oakeley Slate Quarries Co. Ltd., for supplying the whole of their electrical requirements at the Oakeley Quarries from Scheme (a) alone, and it would have been possible to have done this at ½d. per unit, which was very substantially below what was being paid at the time to the North Wales Power Co. Moreover, with the acquisition of the Votty and Bowydd Quarries, and their interest in the Yale Power Station, and also by an arrangement to supply the Rhosydd Quarry with electric power instead of water power, the water from Conglog Lake and other sources supplying the Rhosydd Quarry could be taken and stored in the Cwm Foel reservoir, thus doubling the output for both schemes, exceeding a total of ten million units per annum.

To give an idea what the output of these schemes represent in cash, ten million units would come to £20,833. However, for reasons which I need not state, but which did not include any engineering difficulties, the schemes were not proceeded with.

Blaencwm was built as a hydro-electric station to power Croesor Quarry (Chris Andrews Collection)

31. ON THE COUNCIL

Up to the beginning of the last century, popular education was non-existent, and especially in rural districts the great bulk of the population could neither read nor write, were in profound ignorance of what prevailed in the outside world, and were absolutely unable to take part in local government.

Generally, a large country land-owner, owning perhaps nearly all of the parish was the only one possessing education and experience necessary to direct local government. The dominating power being land-owning and agricultural, the affairs of the Parish were administered in a way to keep the burden of the rates upon agricultural land as low as possible, with a minimum regard for public convenience.

For the majority of people, power and the ability to control the destinies of other people is very attractive, and once possessed is not readily relinquished. The landowning class, who were also the governing class, were ultra-conservative and unprogressive in their political views. Progressive ideas cost money, and they were liable to be touched in a very tender spot, i.e., their pockets.

With the coming of a measure of popular education, School Boards, County Schools, District Councils, and Parish Councils came into existence, and the Democratic control of public affairs was encouraged and promoted. The interference and substitution of democratic control for autocratic control that had hitherto prevailed was distasteful to those who had previously exercised it, and for a long time the habit of subservience to the dominating landowner or squire helped to perpetuate his influence.

His influence was also augmented because he belonged to a class from which appointments to the Magisterial Bench were invariably made. There was, moreover, a strong alliance with the powerful vested interest of the Established Church. Nevertheless, perhaps it was the only practicable arrangement under the circumstances, and was accepted with docile acquiescence by the other parishioners.

I am not blaming those who were hitherto considered the ruling class, as they were simply following the instincts of average human beings, and when such innovations as Public Councils were instituted they would naturally try to control these, so as to maintain a policy consistent with their own ideas and interests.

Llanfrothen Parish being rural and agricultural came into this category, and it was a long time before the un-progressive Tory domination gave way to more enlightened and democratic methods.

Throughout my life I have been intolerant of inefficiency, and progressive in my instincts. I have been called by both friends and critics an "efficiency Maniac", which I would certainly prefer to be rather than a "stick-in-the-mud" muddler. Psychologically, I am afraid that my bump of veneration is on the small side, and I am incapable of the almost idolatrous homage which I have seen some people pay to persons holding high positions, simply because that person held that position, and not because he deserved it or was even worthy of it.

This does not imply any antagonism on my part to any such person, but it will be obvious that temperamentally my place would be found among the progressives. I would like, however, to stress the point that the cause separating me from the old and reactionary influence that had dominated the parish, was ideological and not personal, and wish this statement to apply to all my public life, as in the larger sphere of the County Council of which I afterwards became a member I never allowed personal likes to interest me in the slightest degree in the course I pursued.

The gentleman who had been almost the reigning monarch of Llanfrothen Parish I will call Mr X, and his son Mr Ab X. I was told that the way in which Parish Council elections were conducted was that at a Parish Meeting, presided over by Mr X, the Chairman would name persons whom he thought suitable for Councillors, that no-one dared oppose his nominations, and that no contested election where the ballot box would have given the opportunity for expressing a real opinion had ever been held, and that the same thing had occurred in connection with the District Council and the County Council, for both of which Mr Ab X was nominated. No one had the temerity to put forward any other candidate, much less appoint another in place of Mr Ab X.

The result was that as regards the Parish Council, its function remained inoperative, and the Chairman would congratulate the members of the Council at the end of the year upon the pleasant association that had existed between them during the year, and that in particular they had had to spend no money.

As regards the other Councils, a minimum of benefit was derived, while of course the Parish could not escape its quota of contribution. I have already alluded to the very unsatisfactory state of communications between Croesor and the outside world, and I made it my business to see the agent who was also the solicitor of the Park Estate, to ascertain what could be done to rectify the anomalous condition in which matters stood.

The road leading from Tanlan up to the Park Gate (Llidiart Budr) about 1½ miles in length, originally constructed as a private road, but used without let or hindrance for upwards of twenty years as a public road, was badly out of repair, and the agent told me that the Estate had no objection to its continued use by the public provided it was repaired at the public expense.

That was before the time when heavy expense on road repair in country districts was even dreamt of, and all the repair that was contemplated was something between £15 and £20 for filling in the ruts. I wrote a letter to the Deudraeth District Council suggesting that this repair should be done, and that the road should be taken over as a public road (with or without the formal consent of the landlord), meaning thereby without a formal dedication of the road to the public service.

This formality was not a legal necessity as the public had already acquired a right to it by use, and I already knew that the landlord would not oppose such action. I sent a copy of the application to the Parish Council asking for its support, but apparently the Chairman misunderstood and was dreadfully shocked at the suggestion that the road be taken over with or without the formal consent of the landowner. The result was that an obedient Council passed a resolution opposing instead of supporting the application.

Application was also made by a deputation from Croesor to Ab X, the Llanfrothen representative on the District Council. Llanfrothen's contribution to the District Rate was two seventeenths of the total, so that if seventeen pounds had been spent on the repair of the road only two pounds would have been payable by the Llanfrothen Parish.

I have the authority of the Chairman of the District Council—Mr William Jones, Penybryn, a gentleman for who I had a great respect—for saying that the whole of the District Council with the exception of Ab X were in favour of carrying out the work, but he said: "What can the Council do in face of the opposition of the local member, and of the Parish Council!" The result was that due either to extreme prejudice or unwillingness to spend even two pounds on behalf of the Parish on such necessary work, the opportunity was lost and nothing was done.

This arbitrary action was strong evidence that so far as local Government was concerned the Parish was practically disenfranchised, and to me the position was intolerable. I called a public meeting, put the case before it, and suggested that whatever the outcome of the elections as regarded the voting, it was advisable to nominate candidates who would be prepared to contest the forthcoming elections of County, District and Parish, even if only as a protest.

No one could be found who would stand for the County Council election, although I pledged myself to give all the support in my power to any one who would. I certainly had no desire myself to become a member of the County Council, as I had so many other things demanding my attention at the time, but failing the finding of any other candidate I decided to contest the County Council election myself, and to my intense surprise was elected by a majority of ten votes over the sitting member.

A Committee had been appointed by the Public Meeting to select candidates for the district and parish councils, and this was done. Whether what followed was the result of neglect or design I cannot say, but on going home from my office one Monday, a neighbour informed me, as I was passing along the road, that a Parish Meeting was being held that same night at the Llanfrothen Schoolroom, to elect members for the Parish Council! "Why, then", I asked, "have no notices been put up to that effect?" "There was one put up last evening, Mr Kellow," he said, "in the porch of the Croesor Chapel, but no posters were put up as far as I am aware." I immediately went to the Croesor Chapel, read the notice, and observed that it was dated the 7th day before the meeting, whereas the Act of Parliament specified seven clear days, which meant that the notice should have been dated at least eight days before the meeting.

Capel Croesor Chapel (David Medcalf Collection)

I took possession of this notice, and without waiting to get any dinner, got four other men to join me and we went as quickly as we could to the Llanfrothen Schoolroom.

We found, as we expected, X in the chair, and when I came in I asked him what had been done. He said that the election was all over and the members duly elected. I protested against the legality of the proceedings, saying that the Meeting had not been properly advertised and that, therefore, we had been deprived of the opportunity of nominating candidates of our own choice. I asked him to give us permission to nominate our candidates, and take the vote of the Meeting upon them and any other candidates who might be nominated.

He declined to do this, and I then produced the Notice that had been posted in Croesor Chapel, showing him that even if they had advertised, they were illegal because they did not give the required 7 days notice.

I said that I would not press this point if he would agree to re-open the election, but he again refused. He then got up and went home. I immediately went to the Chair, wrote out a formal protest against the legality of the proceedings, signed it myself and got a number of others who were there present to sign it, and sent it after him to his home the same evening.

The Returning Officer was Clerk of the Guardians and of the District Council and a strong Conservative. Very shortly afterwards I went to see him and asked him what had happened about the Parish Council election. He said that the Council had been duly elected. I told him what had happened, and of the protest sent to the Chairman.

He answered that he had nothing to do with it, as the Chairman's certificate that the Council had been duly elected was sufficient for him. I did not feel inclined to be baulked in this way, and got up a petition to the County Council to intervene and quash the election. This they did, and another Parish meeting was called to elect Councillors.

The Committee now referred to saw to it that their candidates were nominated not only for the Parish Council but for the District Council, and in the result I think I am correct in saying that not one single member of those nominated at the previous meeting, or had sat in the previous Council, was elected, and Ab X lost his seat on the District Council as well as on the County Council.

One would have thought that after this Croesor would have got its road, but we found that we were not at the end of our troubles. The Returning Officer before mentioned, and his very active and energetic assistant became our deadly enemies.

It is remarkable that this Returning Officer was the Worshipful Master of the Madoc Lodge of Freemasons during the year in which I was initiated, but the Brotherhood of Freemasonry did not much lessen the acerbity of his opposition in the public sphere.

I have already said that he was Clerk to the Guardians, of which body X was a member, and I think Chairman. He was also Clerk to the District Council, and it will be readily understood how, with officialdom working against one, not much progress could be made. Later, when this man died, the position was given to his former assistant, who was more rabid in his opposition than his predecessor had been.

I rather think that this opposition was due to my refusal to accept his ruling on points of law relating to rating, and of course any suggestion of fallibility on matters relating to his office was unpardonable. Notwithstanding the unfavourable atmosphere created in the District Council, signed petitions, resolutions of public meetings, and deputations were sent to the District Council, and at last this council itself convened a Parish meeting to take a vote of the ratepayers on the question.

This meeting was addressed by the Clerk who, had he done his duty, would have put the pros and cons fairly before the parishioners, but instead it was one of the most violent ex-parte speeches I have ever listened to, raking up almost everything that could be said against making the road from Tanlan to Croesor. The schoolroom was packed almost to suffocation, and despite the extremely strong opposition of the Clerk, a vote in favour of making the road was passed with only six dissentients, the latter consisting of Ab X and those closely associated with him.

A deputation was also appointed to attend the next meeting of the district council and urge upon it the desirability of making the road forthwith. The Clerk then jumped up and said that it would be better if the deputation included one member from the minority, and suggested adding Ab X to the deputation.

The Meeting did not appreciate the motive of this plausible trick, and when the deputation, which included myself, appeared before the District Council we noticed that Ab X was sitting some distance away by himself. I got up and said to the Chairman that Ab X had been appointed a member of our own deputation to ask the Council to make the road, and asked why, if he was there in this capacity, he was not sitting with us.

The fact was that the Clerk had so manoeuvred matters that he intended Ab X to speak not in favour of, but against the making of the road, and sensing this I asked that he be not heard, as he had no *locus standi*. One of the Council made a remark that he had a right to be present as a member of the public, and I replied that if he were allowed to speak it would have to be understood that it would be as a member of the public, and not as a deputy of the Parish Meeting.

However, he was allowed to speak, but our protest had been registered. But in spite of all this, nothing was done.

The next phase was that the District Council roads were taken over by the County Council, and I was in hope that in a body less open to local influence and prejudice something would be accomplished. Several public meetings were held at Croesor at which our local representative on the County Council was invited to be present, but he never came, nor did he ever take any notice of resolutions passed which he, as the representative of the Parish was asked to support.

I could not understand this apathy, as it never dawned on me that he could be opposed to the making of the road. He was a member of our District Council as well as the County Council.

He was also Clerk of the Parish Council and Assistant Overseer. To my surprise, a ratepayer in Llanfrothen told me that he was an opponent and not a supporter of the road, and gave as a reason for this the fact that the man was son-in-law of the Clerk of the Parish Council who had made the bloomer on the notice of the Parish Meeting, to which allusion has already been made.

This man had been a strong ally and friend of X, and there is very little doubt that he felt the humiliation of the serious consequences which were the result of that mistake. It appeared that the resentment against Croesor and against what I stood for even extended to the "in-laws" and we were back once more in the position—that our elected representative on County and District Councils, instead of supporting the policy approved by the Parish, was doing all he could to prevent its being adopted. Well, I have heard it said that "while a man is the head of the house, the woman is the neck, and many a noble head has been swayed to one side or the other by the movement of the neck." After the responsibility of the maintenance of the District Roads had been transferred to the County Council I made an application to this body to construct the road from Tanlan to Croesor, and a Committee was appointed to view the place. Mr Sam Evans, the Engineer of the Ministry of Transport, met the Committee on the site, and thoroughly approved of making the road in preference to spending money in tinkering with the old Coed Ty road.

I may here interpolate that at that time the great cloudburst occurred over Merionethshire which damaged the Dolgelley Bridge, causing the river to overflow its banks, and inundating the town with sewage matter. A rubbish heap at the Park Quarry which had been tipped across the river by previous owners and under management antecedent to my own, was broken away and carried down into the river bed below, thereby filling it up and causing the river to break away the road for some distance.

This damage made the scheme look much less attractive than it otherwise would have done, but in no way precluded the making of a satisfactory road at a somewhat higher level just opposite the short distance along which the damage had been done.

If, however, there was any question of any further damage being done from the same cause at a later date, there was an alternative route in the field at the edge of the Ceunant (gorge) which was perfectly graded and could be adapted with the landowner's consent.

The whole situation created by the damaged road was fully observed by Mr Sam Evans, and the County Council Committee, and Mr Evans approved of the site and a statement that the Ministry of Transport would be prepared to contribute 50% of the cost was made with all the facts before him.

The site had also previously been inspected and approved by the County Surveyor. A responsible firm of Road Contractors had offered to make a road, up to the standard for the average District Road, from Tanlan to Park Gate (Llidiart Budr) with slate debris, for a figure slightly under £2000, but County Council standards were higher than District Council standards, and granite instead of slate debris was favoured as the material of construction, and £8000 was named as the probable cost under these conditions.

This figure though substantial looks small in comparison with present day expenditure on the public highways of the country, and the opponents of this scheme by this time augmented on the County Council by the most talkative of the members of the District Council, tried to make out that £8000 was too much to spend on making a road to a small country village like Croesor, and decision was again delayed on this account.

I may here remark that in my contact with human beings in general, and not excluding that august body the County Council, I have found that the majority have no real conception of what large figures mean. Their experience in life has been limited to the consideration of their weekly wages or monthly salary, the turnover of their small shop, or some other small amount, and when larger sums are involved their real conception of amount say over £100 are very similar to those conveyed by a thousand or even a million.

There is a vague idea that some large amount of money is involved, but an accurate understanding of its magnitude does not truly exist. For this reason, I have known members who I regarded as intelligent and capable when dealing with ordinary affairs, incapable of grasping figures such as are encountered in the financial operations of the County Council, and of the incidence and interactions of the various factors involved in the Income and Expenditure of the County Council.

A profound change in the management of the Finance of the County was brought about by the de-rating Act. Prior to the passing of this Act, a penny rate yielded over £900. I do not recollect the exact figure, but I know it was nearer to a thousand than nine hundred. By the underating of a large number of properties in the County, a penny rate on the remainder produced only something over £500, and this gap in the Income of the County had to be made good from some other source, if the services of the County were to be maintained.

There were some minor changes in the manner of our dealing with grants which accrued to the County, but the deficiency in the amount yielded by the rates under the old and new conditions was made up by the Treasury by a lump-sum payment known as the Block Grant.

The amount of this block grant was varied from time to time according to the needs of the County, and at first a triennial period, then a quadrennial, and finally a quinquennial period was adopted for making adjustments in the block grant.

One year was elected out of the period for ascertaining the amount of the deficiency that had to be made up, and the amount so ascertained would be paid for the whole of the succeeding period. The year selected for making the calculation was always the penultimate year, thus in a quinquennial period it would be the fourth year, the object no doubt being to await the completion and auditing of the County Accounts during the fifth year, and the determination of the amount of the block grant prior to the commencement of the succeeding quinquennial period.

By way of minimising the objection to spending £8,000 on a road to Croesor, I made the statement to the Council that over a long period of years the County had only to contribute an average of £273 out of every £1,000 spent on the main roads, and this was based on a 50% grant from the Road Board, which reduced the £1,000 to £500, and by the Block Grant which paid nearly half of the remainder.

To my great surprise I was interrupted by a prominent member of the County Council, and whose position should have enabled him to speak authoritatively on the point, by the remark; "We cannot use the Block Grant for the main roads," and a few minutes later another equally prominent member of the Council, from whom one would ordinarily expect a knowledge of the legal position reiterated the same remark that the Block Grant was not available for the main roads.

Rates both before and after the passing of the de-rating Act had been levied and spent on the making and repairing of the roads, and how the deficiency created by the de-rating Act was to be made up except by the Block Grant, in precisely the same way as such deficiencies were made up for every other County purpose, was a point I would have liked these gentlemen to explain.

As a matter of fact, in the consideration of the budget figures there was a separate estimated allocation for every item of expenditure in the County but the total so arrived at without distinction of the items were where its component parts became the basis of the amount it was necessary to provide partly by Block Grant and partly from rates.

The Income, though based on an itemised budget and consisting of ordinary grants, block grants, and rates were all paid into a common fund called the County Fund, which was drawn upon indiscriminately for all the County purposes. It is, therefore, quite obvious that main road in common with other services participated in Block Grant money.

I have not got the annual accounts for the various years before me as I write, but I do recollect that in one year the total expenditure of the County was substantially £180,000 for the year. Road Board, and grants other than block grants, amounted to £60,000, leaving to be made up by block grants and rates £120,000. The Block Grant was then deducted from this figure, and amounted to over £59,000, leaving something less than £61,000 to be provided from rates.

This was a favourable year, but other years were not widely different. It will be noted that of the amount to be made up by block grant and rates, almost 50% was made up by Block Grant.

While I was endeavouring to get the Council to undertake this work I went to London where I stayed for some time.

During my stay, a possible development of the financial situation occurred to me, and being in London gave me the opportunity of calling at the Office of the County Councils Association.

I explained that I was calling in a private capacity and not on behalf of the Merioneth County Council. Nevertheless, I was very cordially received by their Clerk, who was also a Solicitor, and probably the highest authority in the Kingdom on matters relating to County Councils.

I told him about the proposed road to Croesor, which it was suggested would cost £8,000. It was then at the very beginning of the penultimate year on which the block grants were calculated for the ensuing quinquennial period.

I put it to him that if the road was made and paid for during that penultimate year, £4,000 out of the £8,000 would be paid for by the Road Board, leaving a deficiency of £4,000 to be made up from rates. But as the deficiency would be taken into account in fixing the amount of the block grant for the following quinquennial period, and assuming that the additional block grant would be roughly half this £4,000 deficit equal to £2,000, the Council would ultimately collect £2,000 for each of the years in the following quinquennial period, making a total cash receipt with the £4,000 from the Road Board of £14,000 against the total expenditure on the road of £8,000, thus enabling the Council to pay the whole cost of the road, and make a cash profit of £6,000.

Of course, I was only speaking in round numbers, but they were not very wide of the mark.

The Clerk replied that I had made a very interesting point, but thought it likely that I was right in my forecast. He kindly promised to look up the law on the matter, and write me his considered opinion. This he did, and it was to confirm the soundness of the principle of my forecast.

Notwithstanding this authoritative opinion, I failed to get the Council to move, and poor unfortunate Croesor was again denied its undoubted right to have a proper road made that would give it access to the outside world, such access being only obtainable by a road making Tanlan its starting point.

Large investments of capital in the locality, with consequent industrial expansion, would undoubtedly have taken place if this condition had been fulfilled. The money expended on making the road would have been spent in the locality and benefitted not only the workmen but the tradespeople and other members of the community. But notwithstanding resolution after resolution from Croesor, resolutions by the Parish Council, and a nearly unanimous vote by a Parish Meeting, a deputation representing the whole Parish, a petition signed by nearly every parishioner, and by doctors in the vicinity, in addition to the favourable financial position already referred to, a resolution refusing the application—which was tantamount to a death warrant to a Welsh village in the County—was passed by the County Council. When one considers these additional facts, namely (a) the landlord of the Park and Croesor Estates offered to contribute the existing road, (b) any additional land required to complete the road, (c) The use of the excellent material included in a half-mile long massive granite wall, (d) the right to supplement this by quarrying granite at the Garreghylldrem Granite Quarry, (e) the use of slate debris wherever it was situated—all free, gratis and for nothing, one wonders if the wire-pullers who brought about the deplorable result will ever have enlightenment, and regret their unfortunate action.

When first I became a member of the County Council, I was elected a member of the Main Roads Committee and of the Finance Committee. Very shortly afterwards I was elected Vice-Chairman of the Main Roads Committee and from that position naturally stepped into the Chairmanship in the ensuing year.

This was exceptional and is the only instance I am aware of where a Vice-Chairman of a Committee was ever appointed. I was also at a very early stage in my career as a Councillor, appointed Chairman of the Finance Committee, which position I held continuously for eight years, and then retired as I had been elected Chairman of the County Council itself, and could not hold the two Chairmanships at the same time.

I possessed expert knowledge of Advanced Accountancy by Double Entry, including preparation of Profit and Loss Accounts, Balance Sheets, and other accounts prepared to enable the owners of the business periodically to know their financial position, and the factors contributing to the gain or loss resulting from the business.

I certainly expected in connection with a body holding so important a position in the country as the County Council that something superlatively good in its system of Accountancy would be in use, which would enable one to take a comprehensive and accurate view of the financial position of the County, so that intelligent management could be exercised.

I was astounded to find that the contrary was the case. To the ordinary business man it would be incredible that a business such as that of a County Council, where large sums were involved, should be carried without a single Ledger Account to show the position of any Debtor or Creditor in relation to the County Council, yet this was undoubtedly true.

If, for example, granite was bought from a Granite Quarry, costing, say £700, or certificates were issued by the County Surveyor for £1200, for work carried out on the roads, or on the other hand that precepts for rates for rates payable to the County Council—not one figure or letter was ever entered in any book or record to show the transaction, and the result was that the Council never knew its Assets or its Liabilities.

The conception that when, say, a Granite Company supplied the granite and sent a monied invoice it then became a Creditor of the Council, and that the Council simultaneously became its Debtor, and should have had not only books of prime entry, but a Ledger Account in which the particular department of the County Council concerned (such as Main Roads), was debited, appeared to be beyond the capacity of the great craniums who conceived County Council accounts.

Centuries ago the fallacious idea was taught by political economists that wealth consisted of money, and property other than cash was not admitted to be wealth. One of the consequences of this was that it was held that trade carried on with a foreign country either increased or decreased the wealth of this country, according to whether cash came into the country or went out, regardless of the fact as to whether value in any other form was imported, or exported.

Subsequent writers on political economy such as Adam Smith, and John Stuart Mill have expressed surprise that so obvious a fallacy could ever have been adopted by the nation, but make the remark that error universally accepted is believed and acted upon by intelligent men because of the universality of the belief.

Now, it will be believed that a survival of this erroneous conception profoundly influences the national, local government systems of accounts as also the accountancy of our Law Courts, even up to the highest, and consequently in the system of accounts prescribed for County Councils by the Local Government Board, nothing is considered as having happened in the way of a business transaction unless and until cash has been paid or received.

Therefore, the first and only entry that would ever appear in the County books in connection with the purchase of the granite referred to above, would be the entry in the Cash Book of the payment on the date on which the granite was paid for, possibly months after the granite had been laid on the roads.

As Chairman of the Finance Committee I told the members that although I was Chairman I had no idea how the County stood financially, and that it was certain that no member of the Committee nor any other official had any idea either, and went on to explain why.

It was decided that I should interview the Auditors, which I did, but was confronted by the statement that the existing form of accounts had been prescribed by the Local Government Board, and that they could accept no other.

I pointed out the unsatisfactory state of things that resulted from this and was told that the Council could keep any other books that it chose for its own information, but that it would be outside of the ambit of their duties to take any official notice of them.

I reported this to the Committee, and advocated that we should keep a set of books founded on debits and credits in addition to the set of books founded on "Cash Paid," and "Cash Received." This was approved in principle and the County Treasurer agreed thereafter to produce an account quarterly, to be called the "Income and Expenditure Account," based on the Credits and Debits of the preceding quarter, and on the presumption that an initial balance sheet of the liquid assets and liabilities had been prepared, a statement at the end of each quarterly account of the excess of liabilities over assets or of assets over liabilities. By comparing with preceding quarterly statements it could then be seen at a glance whether the financial position of the County was improving or deteriorating. This appeared to have satisfactory results, but I am afraid that the Treasurer was not a convinced man, and knowing that the accounts as he had hitherto kept them satisfied the requirements of the Local Government Board, regarded a second set of accounts showing "Income and Expenditure" based on Debits and Credits as a redundancy, and although thereafter the quarterly accounts included what purported to be an "Income and Expenditure Account" I do not think that it was ever backed up by regular books of prime entry, journalised and entered in a Ledger.

For this reason I was never satisfied with the reliability of the Income and Expenditure Account, and I am afraid that my constant criticism caused the County Council Treasurer to think that I had a personal grudge against him, which I certainly had not. I regarded him as absolutely conscientious, a very hard-working official, his books on the cash account system were a model of neatness, and he would go to any amount of trouble to supply a member of the Council with information on any point. His salary was very low, and if he could only have been induced to keep the Income and Expenditure Account in the orthodox way, I would gladly have voted for doubling his salary.

But a man who has been in a particular groove for forty years does not readily change it into another. That is just ordinary human nature, so while expressing regret at what was taking place, I did not and do not express surprise.

I am not at all sure that even a large number on the County Council thought that I was faddy and some men intelligent in other things went so far as to say publicly "of course there can be no expenditure until cash is paid." After about 25 to 30 years, however, I obtained my vindication. The Auditors' report on the Annual Accounts was being read to the Finance Committee and for the first time it condemned the keeping of the account on the Cash Account basis, and said that it should be kept on the basis of Debits and Credits. In fact, the whole report was in its entirety what I had been advocating for such a long period of time, and the member of the Council who was reading the report finished up by saying: "signed Moses Kellow!" Of course, it was actually signed by the Auditor, but the introduction of my name was a jocular way of admitting that it exactly coincided with what I had always advocated. During the later years of my membership of the Council, I was suffering from the effects of a series of serious operations (to which I have alluded earlier) and was in a low state of health. In addition to this, I had serious business responsibilities and it became difficult for me to be regular in my attendances at the meetings of the County Council and of its Committees.

For this reason I am unable exactly to say what happened with regard to the accounts after the report of the Auditor to which I have referred. While on the topic of what is known as the Government system of cash accounts, I may say that once, during a visit to London, I met a Director of a large electrical manufacturing company, and he said to me: "Mr Kellow, do you know how the Government keep their accounts?" I said that I did not know for certain, but probably they were kept on the same system as County Council accounts were kept, with which I was familiar.

"Do you know," he then continued, "that they have no idea how much money they owe us, and if we say £80,000, they send us a remittance for that amount, and if we were to say double that sum we would get a remittance for that just as matter-of-factly. It is a good job they are in honest hands." Here again, it seems almost certain that no proper Ledger Accounts were kept. "We have heard again and again of accounts being paid twice over," continued my friend the director, "by the Government in respect of the same indebtedness, and I myself recently heard of a well authenticated case where not only was the firm paid twice over, but when the money was sent back, giving dates of the previous payment and the number of the receipt, the money was returned to the firm, with the denial that the account had been paid twice over." The amount involved in this transaction was between £800 and £900. In another case where an account had been paid twice over the professional accountant of the firm concerned went to London to the Government Office to refund the money, and he told me what occurred at the interview. Expecting to be thanked for the return of the money, he was angrily told that no such thing as double payment for one indebtedness was possible, and that the Office were always being bothered in this respect by "little" provincial firms. Big firms never bothered them in this way, and were far more satisfactory to deal with.

It was quite evident that this Official would far rather the Government lose the money than that he should have to make a record in his books of the receipt of a refund. No doubt the bigger firms had had a wider experience of the workings of Government departments, and concluded that when a beneficent providence, through the agency of a Government Official, put redundant money from the National Exchequer into their pockets, it was not for them to carp or in any way thwart the inscrutable ways of providence. Besides, they would not be so inconsiderate as to embarrass the Government Department by pointing out that one of their Officials was capable of paying twice over for goods or services rendered to the Nation.

One hears a great deal about the Millennium that is going to occur after this great war, and how presumably from the enormous debt that has been piled up, vast riches will accrue which can be distributed with a free hand to the deserving and undeserving alike through the medium of Government Departments, who will control everything, even to the number of buttons to be put on a pyjama jacket. And this miracle will be performed by the fine business management exemplified in the cases cited above.

32. TALK ON AGRICULTURE

During the year in which I was Chairman of the County Council I was invited to preside at a meeting of Women's Institutes at Barmouth, called to do honour to Miss Talbot, the Head of the Women's Department of the Board of Agriculture. Miss Talbot was to give a half-hour's address, and I was asked to follow with an address on "Rural Needs." When I received the request, I treated it very lightly, thinking it was a matter of comparatively local importance, but all the Presidents of the Women's Institutes of North Wales turned up at the meeting, intelligent women determined to be handed real information, and not foisted off with easy platitudes. More than five hundred women were present.

In the course of my life I have many times been in hazardous positions, but these were nothing to the risk I ran when at Barmouth, I, as the only male in the room, faced these five hundred determined women. However, I survived the ordeal, as although I had paid little attention to the preparation of my address on rural needs, I decided to make my theme the desirability of promoting efficiency in rural affairs, and in particular with regard to agriculture and farming.

I told them that a great friend of mine, Sir William Arbuckle, who was the Agent-General of Natal Colony in this country, frequently visited my home in Croesor. One day while he was there he saw a man ploughing a field, and enquired how much he could do in one day. I replied about half an acre, to which Sir William said: "I have an estate in Natal, and we reckon to do fifty acres a day using a twelve-furrow plough." I implied that that was surely expensive equipment beyond the means of any but large estate farms, but he replied: "True, but the plough is owned by ploughing contractors, and if I want 300 acres ploughed up, I engage them to come to my estate for a week, and when they have finished my job, they move on to another estate, taking their plough and its appurtenances with them." This was more than 25 years ago, when these new marvellous machines such as draincutting and laying machines, combine harvester, etc., were in their experimental, though practical stage. The war has brought them into common use and efficient distribution.

In my address to the Women's Institute members, I advocated the stripping away of many of our hedges, thus giving access to large tracts of land available for mechanical cultivation, and thus promoting more progressive ideas and experimentation in the science of agriculture.

I also impressed the view to my audience that there must always remain the problem of labour. The remuneration of farm labourers, a highly skilled body of men or they are utterly useless for the complicated operations involved in agriculture, must be made high enough to encourage the most intelligent men to take it up as a livelihood. Whatever theorists may say, men will not work when paid day work in the same way as if paid by results, and farming will have to compete on an equal footing with industry.

Probably the best way would be to work out a system of payment by piecework, but if this were impractical, possibly an alternative method would be to run every large farm as a “Joint Stock Company” with the employees as shareholders. I also dwelt on the subject of improving our cattle and sheep strains, and spending money on research into every branch of agricultural activity in order to raise the general standard of efficiency.

I must admit, however, as I aired my own private views to my listeners, that I felt myself something of a fraud for posing as an agricultural expert, but they appeared to think otherwise and I was dumb-founded to be accorded a tremendous ovation. I was overwhelmed by congratulations by as many who could get near me to shake my hand.

I had a private talk with Miss Talbot after the meeting, and made contacts intellectual and personal which I think had repercussions later on. Shortly afterwards, I received an invitation from the Countess of Sandwich to come to her house in London to meet “Mr Protheroe”, the then President of the Board of Agriculture. A little later I received a second invitation to come to her house to meet the American Ambassador, who was much interested in the agricultural activities of this country.

He expressed the view that revolutionary improvements in an Industry did not usually emanate from those actively employed in it, but from poets who had the gift of imagination. He said that in the USA they found it very difficult to get new ideas adopted by the farmers, because they thought they already knew everything there was to be known about their business, and were therefore unteachable. He cited an experience that he had himself had at a place where a meeting had been called by a very responsible scientific body, to acquaint farmers with the result of their recent investigations.

After the meeting, he met on the railway platform a farmer and asked him if he had attended the meeting, at which the main topic expounded was the best manner of feeding pigs. The farmer said: “No, I don’t agree with the scientists. Raising hogs is my job, and the only way to do it is to feed plenty of corn.” “Well,” said the Ambassador to me, “that farmer was wrong. The best way of raising hogs is to feed them on hay.” This was a new idea to me, but it shows, if the scientists were right, that there is often an alternative way of attaining an object to the one that is considered orthodox. On the occasion of these visits to the Countess of Sandwich’s London residence, I met many of the most influential men in the Government, and titled men of culture and conversational brilliance, and altogether found my visits intensely interesting.

33. COUNTY COUNCIL MATTERS – VOTING ON AMENDMENTS

To return to the County Council, I was elected to represent the County Council at the two public enquiries held at Llandudno to examine schemes for the electrification of the North Wales and Chester Electricity Area, and also at the various Conferences held at Llandudno Junction.

At one of the latter, a rather surprising thing happened. Representatives of the various County Councils, Chester Corporation, the North Wales Power Co, and others were there, seated around a table which ran the whole length of the room. Mr Page who was the Vice Chairman of the Electricity Commissioners (and was subsequently Chairman) was in the Chair, and I was seated at the opposite end of the room.

Mr Page got up, walked the whole length of the room to where I was seated, held out his hand and said: "How do you do, Mr Kellow," and then returned to his Chair.

He took no notice of any other person in the room. I had never been introduced to him, nor spoken to him before in my life, and why I should have been selected for such special and marked attention in the presence of so many other important people I do not know.

There is a point in procedure at County Council Meetings and also at meetings held throughout the country, where resolutions are passed on various matters. The usual proceeding is for someone to move a resolution, and if it is seconded it is then before the meeting for discussion. Before the discussion is closed, the Chairman will ask: "Does anyone wish to move an amendment?" and if an amendment is moved and seconded it now comes before the meeting for discussion.

When the discussion has ended, the Chairman will say: "All in favour of the amendment please signify by holding up their right hands." These are counted. He then says: "All in favour of the original motion," and the votes for this are taken in the same way. If more votes are cast for the amendment than for the original motion, the amendment is declared carried, and the original motion is dropped altogether.

I entirely disagree with this method of procedure. One cannot have two separate and distinct resolutions before a meeting at the same time, and the very word amendment implies the existence of something that can be improved or amended, and that something cannot be anything else than the original motion, and should be capable of being incorporated in it.

Any so-called amendment that cannot be incorporated in the original motion is not an amendment at all, and should be disallowed by the Chairman. It might conceivably be that a voter against both original motion and the so-called amendment, might very easily vote for the amendment as being the lesser of two evils.

During my Chairmanship of the County Council itself, I refused to adopt the usual course.

When a legitimate amendment was being voted upon I put it to the meeting: "For the Amendment 'X' votes; against the Amendment 'Y' votes!" If the votes for the amendment exceeded those against, I then declared the amendment carried, but proceeded to say: "The original motion as now amended reads as follows: --------" and then took another voting for or against the amended original motion. Under this method a member might support the amendment, and yet vote against the original motion, in which the amendment had been incorporated. I have seen many members look very puzzled at my way of taking the votes, but no one ever challenged its fairness.

34. THOUGHTS ON RELIGION

I will now deal with a phase of my life in which I have had somewhat unusual experiences, and this relates to religious matters. My readers may think (a) that I am a "goody-goody"; (b) that I am a cosmopolitan; and (c) that I am an unbiased seeker after truth, and I think I may range these in the ascending order of probability.

Between the ages of six and nine I went to the Welsh Chapel at Tremadoc and attended the Welsh Sunday School where I learned my "ABC" and to read Welsh. After the Welsh service was over, I was taken to the English Church across the way, where my father had a pew. Then we came to Croesor, and I attended the Welsh CM Chapel for services and Sunday School, but as my father could not easily follow a Welsh service, he would frequently go to Portmadoc and attend either St John's Church or the English Wesleyan Chapel in the Park. The first being very dignified and aristocratic, where any ejaculations by the congregation would have been unthinkable, to the very proletarian Wesleyan Chapel where "Amens," "Blessed be the Lords" and groans as though they had the belly-ache were the commonplaces of the service.

On these visits to Portmadoc, my father would often take me with him, but as I got older and became very closely associated with the musical part of the services at Croesor, I had a place in the "set fawr" (Deacon's Pew) as a "Codur Canu," and was, of course, a church member, but refused the honour of deaconship.

I dropped the Portmadoc visits. Many people attending the Church of a particular sect will not enter the Church of another sect, but my broadmindedness, thirst for knowledge, or curiosity, call it what you like, impelled me to go anywhere and everywhere where I might get a novel experience, and be able to study new phases of human nature.

As a boy I had the opportunity of attending Moody and Sankey's meetings at Liverpool. I attended services at Spurgeon's Tabernacle on several occasions. I have heard all the preachers at the City Temple, London, from Parker onwards, and on one occasion when RJ Campbell was preaching the Church was so crowded that I had to sit on the pulpit steps. The Rev RJ Campbell afterwards turned over to the Roman Catholics.

I once attended a service that was at St James' Hall, which had been lent for the purpose, where a full orchestra took the place of the usual organ. Two famous preachers, Mr Mark Guy Pearce and the Rev Hugh Price Hughes took the services alternately. Later on, Mr Mark Guy Pearce, who was a Cornishman, a splendid raconteur and very good company, stayed with me at Croesor for a week.

I have attended several meetings at spiritualist churches, one Salvation Army meeting in a large hall in a very populous neighbourhood where restrained conduct was not the order of the day, and at two or three of their great festivals where the Army Chiefs, the Booths, the Higgins, and others were present.

These were of course much more dignified affairs. I have also been in several of the Protestant and provincial and London Cathedrals, including of course Westminster Abbey and St Paul's. I think one of the most extraordinary services I ever attended of the Church of England was at one of the old City Churches.

There were three clergymen, a grand organ, a surpliced choir of not less than thirty members who walked round the church, and throughout the service there was the maximum of ceremony, as this was a High Church—and the total of the congregation, including myself, was six persons! It was in one of those parishes where the tithes and other income were sufficient to provide a flood of money, which enabled the elaborate ceremonial to be carried out, that I witnessed the unusual service for six people.

On another occasion, I was present at a revivalist meeting in a chapel belonging to the United Free Methodist Churches, and the scenes I witnessed there were most extraordinary. Those present being in a frenzy of emotion which amounted sometimes to dementia, gave the proceedings a bizarre atmosphere that affected one strangely. At one part of this service, a man who was in the back pew of the middle sets of seats, jumped up and ran like a hare from the back end of the chapel to the penitent bench in the front, by the pulpit, stepping on the edges of the seat-backs all the way. How he succeeded in doing this without slipping I do not know. On arriving at the front, he threw himself down with great violence on the floor, where he lay writhing as though he were in the extreme of agony. I presume it was an example of sin "biting like an adder, and stinging like a serpent."

When I was returning from my trip to America, the passengers on the boat were almost all Irish lawyers going to Dublin to attend some Conference, and the Catholic service was held on board for their benefit, and I attended it. I also attended a Catholic Service once at the Pro-Cathedral, Dublin, and I was once present at a service at Westminster Catholic Cathedral, London, and heard Cardinal Vaughan preach.

On the continent I attended Catholic services at Cologne Cathedral, and at the great Cathedral of St Peter's at Rome. While in Rome I had the great privilege of attending at the Vatican and being blessed by the Pope. The scenes of religious exaltation which I there witnessed gave me the impression that many who were present felt that they were really in the presence of the Deity Himself.

I suppose that to most people, this should be regarded as the culminating point of one's religious experiences, this genuinely spiritual manifestation. I do not think, nevertheless, that I am ever likely to embrace the Roman Catholic faith, as I am incapable of believing what a Church or any other party tells me I must believe, if it were at variance with what appeared to me to be unreasonable.

Goodness was social, and badness anti-social, and to induce conduct which was social rewards were given and promised, and for what was anti-social punishment immediate and prospective was administered.

A great friend of mine, Sir Frederick Douglas Cunningham, KICE, who ruled a tract of country in India almost as large as Great Britain, and who was one of the most intellectual men I have ever known, took the view that a particular kind of religion that had grown up among a people was the one best suited to its needs and temperament, and that the religion of one country or people could not be transplanted into another country, and among another set of people, without harmful results.

The essential thing in every religion was its association with and encouragement of a high moral standard, and that the doctrines of every religion in the world were ill-founded and probably 99% incorrect. They were the trappings, and the insistence that everyone else in the world should accept our set of doctrines was narrow and bigoted. He said that the country he had governed in India had a standard of morality far superior to what prevailed in this country. The people were honest, clean living, and such a thing as an illegitimate birth was unknown.

Now it must not be assumed that Sir Frederick was an atheist, or not a good Christian. He was indeed a devout member of the Church of England, and I put his point of view forward, as it is a very interesting one.

Religion is an instinct very deeply rooted in Human Nature, and has little to do with intellect. It is world-wide and there is nothing too absurd or ridiculous which will not be accepted and believed by the various races of mankind if the glamour of religion is thrown over it. It probably originated millions of years ago in the process of evolution, before man had attained the status of a human being, and was transmitted downward through all the generations that followed.

The herd instinct which prompted animals of a particular kind to combine for their mutual protection and survival, necessitated respect for and obedience to a leader. A herd of horses will be dominated by a powerful stallion, a herd of cows by a fierce bull, and in the later stages of evolution human beings associated themselves together as tribes, dominated by a chief, and later nations by a king. The homage paid to and subservience to the chief or king, who was often credited with attributes which he could not possible possess, was a species of worship which had its beneficent effect in binding together elements and individuals which would not otherwise have co-operated.

Growing up in this way over the wide area of the whole globe, during a time when intercommunication was impossible, naturally caused a great diversity both in the objects of adulation and in the forms of worship, and worship it was, which the objects of adoration evoked.

Thus there came into being a large number of different religions having, however, one thing in common, beliefs in miraculous happenings, supernatural occurrences, and the exaltation of the object of worship. Such I believe to be the state of things in the world today. In religions other than our own, we find reverence, a sense of religious duty and rigid observance of the rites appertaining to their particular kind of religion.

Taking the origin of all Religions to be what I have postulated, the conduct of the individuals would be classed as either good or bad, the good being what was regarded as beneficial to the tribe or nation, and the bad regarded as what was harmful to the nation.

Affecting the desirability or otherwise of missionary effort, it is stated that in the Palace of the Church of England Bishop of Calcutta, not a single so-called converted Indian servant is ever engaged, as they have a much deteriorated standard of morality and integrity compared with that possessed by the unconverted section of the community.

Sir William and Lady Arbuckle and others have assured me that the same holds good among the Kaffirs of South Africa. Whether the view put forward is right or wrong, I am perfectly certain that the members of the Christian Church are not ripe for its acceptance, and would be reluctant to admit that their tenets are not the be-all and end-all of religion throughout the world. The last thing that any man will thank you for is to separate him from his delusions. In common with the priesthoods of every religion, the priesthood of the Christian religion teach that the cardinal sin is unbelief, and unbelief means non-acceptance by the members of their church of the doctrines and practices they stand for, which I have already premised are not actually supported by logic.

Unquestioned acceptance of and belief in and obedience to their teachings gives them great power and promotes a flow of £sd into the coffers of the church. After all, they are human beings, and tend to influence their own emoluments beneficially, being actuated by motives similar to those which activate the majority of their fellow human beings.

Consider for a moment the claims put forward by the Catholic Church and the Church of England, to have authority to consign a man to eternal punishment in the spiritual world by excommunicating him from membership of their respective Churches—and I am not sure that even the less strict of those non-conformist bodies do not regard withdrawal from Church membership as tantamount to banishment to the nether regions. Consider also the absurdities of canonisation, where the Church allocates to itself the right and power to elevate a man dead for centuries to a place of honour in Heaven. Then again there is the appointment of Patron Saints—St David in Wales, St George for England, St Andrew for Scotland, and St Patrick for Ireland, and at a slate quarry in Luxembourg I saw a life-size figure of the Patron Saint of the Quarry! I am told that among the negroes of the Southern States of America, their idea of the Deity is an old gentleman wearing a frock coat and a silk hat. Now we can imagine half-civilised negroes nudging this old gentleman's elbow to remind him of something he is forgetting, and which he should attend to, but is it consistent with the Majesty of the Deity as we, an enlightened people, conceive him? That he should figuratively speaking require the same kind of reminder in order to get him to do the right and proper thing is indicated by the need for Patron Saints, for that is exactly what these special Saints are supposed to be doing. Thus, for example, the needs of the Slate Quarry in Luxembourg would be entirely forgotten, and it would get bad rock instead of good rock, if the owners had neglected to appoint a Patron Saint to represent their interest with the Almighty.

Alongside such absurdities as these, you will find an intolerance of anything the priesthoods do not consider is covered by what they call "revealed religion". They do not appear to realise that revelation is not all contained in a book, but that in the geological record so rich in the information it gives about the world gone by, and the way it came into its present condition, and which particularly in the Snowdonian district is spread out before one's eyes, and is a revelation comparable with any included in their pre-conceived ideas. The same may, of course, be said of what may be observed by the aid of the astronomical telescopes in stellar regions.

Yet, I have often heard it said from the pulpit that science is a delusion, and have heard violent denunciations of men of learning, earnest and consistent seekers after sober truth, men who possess knowledge of revelation far beyond that of their incompetent critics.

I would like to dwell briefly on the subject of the structure of the universe. The vastness of space is illustrated by the fact that the Milky Way consists of myriads of stars so far distant, the majority of them, that light, travelling at 186,000 miles a second, takes centuries to reach the earth—and that is distance in one direction only! But these are our nearest stars, as out own solar system is but a part of this Milky Way, and those distant stars are in reality part of the same nebula in which we have existence. The largest astronomical telescopes turned in any direction have so far revealed no limit to the stellar systems, and stars contained within the range of these telescopes, and made visible by their being clustered in nebulae, out-numbered the grains of sand on all the seashores of our globe, which, in comparison with outer space would represent but an electron in a molecule contained in the smallest of those grains of sand.

Coming to the Earth itself, the great Greek philosopher, Democrates, evolved the atomic theory of matter about two thousand years ago. He took the view that matter was ultimately divisible into exceedingly small particles which he called atoms, and that these atoms were solid throughout.

This view held sway until comparatively recently, when the great English scientist, Rutherford, made the startling discovery that the atom, instead of consisting of solid matter through and through, consisted mostly of voids; that in the centre of the atom there was a nucleus called a proton, which was positively charged electrically, and that around this proton in one or more orbits electrons which were negatively charged revolved around at a terrific rate of speed, something like one thousand miles per hour; that in structure, both as regards movement and proportion of voids to solid, it very much resembled the solar system.

Eddington, another great scientist, investigated the proportion of voids to solid in the atom, and found that in the vast number of millions of atoms comprising the adult human body, the whole of the protons and electrons, if collected together, and their voids eliminated, would be represented as solid matter by a particle hardly discernible except by the aid of a magnifying glass.

Inter alia, this insignificant speck is all the material part of man, who regards himself as the Lord of Creation, for whom heaven and earth were brought especially into existence, and that he is the pivot of a cosmogony covering the whole universe, and that there has been special divine intervention on his behalf.

As a corrective to his colossal egotism, it should be a salutary practice to hear himself declaimed from the pulpit, and to himself confess that he is nothing but a miserable sinner.

However, to return to our main subject, by very advanced mathematics Einstein evolved the theory of relativity, the truth of which is generally accepted by the scientific world, and in which no exact values can be ascribed to any of what was previously regarded as standard quantities or dimensions, but that they are all variables.

He first amalgamated space and time into one plane, called space-time, and later through, or with the aid of Riemannian Geometry he was able to add a third element to his plane, namely gravitational and electro-magnetic fields.

Adopting his theory, let it be assumed that a rod in relation to a particular object at negligibly low velocity, would measure one yard. At much higher velocity it would become reduced to half a yard, and at a velocity of 186,000 miles a second (the speed of light) it would cease to have any length at all! At the same time, its weight would have increased to infinity. Such variations as these in what we formerly regarded as the constants of matter, must profoundly modify our ideas with regard to the immutability of matter, and cause us to ask ourselves if we really know what we mean by matter and the material universe.

Then again, with relativity applied to time, a twenty-four hour day could be shortened to twelve hours, to six, and at a velocity of 186,000 miles a second, would disappear altogether, so that past, present, and future would be all on the same plane. I confess one's mind reels at such propositions, but if past, present and future are all brought on the same plane, the writing of history and prophecy would be much simplified; that it, again assuming that a historian and prophet could conceivably exist! Until comparatively recently, two propositions have been accepted as axiomatic: (1) the indestructibility of matter, and (2) the indestructibility of energy. But it has now been discovered that matter can disappear and absolutely cease to exist as matter, and that, therefore, material objects composing the material world can be extinguished in their entirety.

Concurrently, energy will be produced, in a definite ratio to the matter that has disappeared, and this phenomenon can be reversed, and energy can reappear as matter.

To me, this fact is one of overwhelming importance in its implications, as it removes the boundary line between the tangible and the intangible.

Again we have the unquestionable existence of telepathy, whereby thoughts existing in one person's mind may be transmitted to another person's mind without spoken word or any of the ordinary means of communication being available, and wherein distance does not seem to have any bearing.

In the course of my life I have had thousands of experiences of this taking place, and as I have stated, telepathy is not limited in its range, but can operate just as easily when the two persons concerned are great distances apart. As one example of this, I have a cousin who lives at Stockport, and at long intervals not pre-arranged, we write to one another. It is almost invariable that when we do write, out letters cross in the post.

What is the medium of transmission of thought! Is it etheric, or is there a medium other than the ether, which may become the vehicle of thought? The ether, though recognised as being capable of transmitting radio waves, which waves are suggestive of light itself, as the length of these radio waves multiplied by their periodicity per second, no matter how these may vary as regards each of the factors, always have a product of 186,000 miles per second. Yet, this ether of space is so intangible that the earth and all other bodies throughout the universe can pass through it without encountering any resistance.

However tenuous the ether may be, experiments have failed to indicate that there is any ether draught as the earth rushes through it at high velocity. If, however, the ether is a dense substance, according to Sir Oliver Lodge's view, the absence of friction is even more remarkable, Now the implications of all these things is to me very intriguing, as it appears to be breaking down the distinction between what we have been regarding as the material world and what we have called the spiritual world. It involves no disrespect to the latter to consider the material and spiritual world as part of one great concept in the universe. I would not presume to put forward such an idea as a definitely accepted opinion, but rather as a line of thought and investigation.

I feel that we are only at the very threshold of knowledge with regard to these great problems, and it makes me feel a very humble person, very disinclined to be dogmatic, and very tolerant of the views of my fellow men. Notwithstanding the fact that I think that they are mostly wrong, I include myself in the same category.

35. HEALTH ISSUES

The years of 1926 and 1927 were prolific of business worries, and I began acutely to feel the need of the companionship and sympathy of a wife. The result was that I married a second time. My wife was the daughter of Alfred Thomas Woodhead who was a landowner, the proprietor of yarn spinning mills and of dye-works at Meltham, near Huddersfield. The marriage took place at Doncaster, where my intended wife had been staying with her life-long friend and bridesmaid Miss Martha Wyer, and my best man was my cousin Mr Josiah Gayer, who was engaged in the wholesale drapery trade in Manchester.

The event which followed fully justified the course I had taken. In the middle of November, 1929, I went to Derby to attend a business interview on the following morning, a Saturday.

After this interview, I had planned to go to Llandudno for the weekend, returning home to Croesor on the Monday.

Actually I did not return home until July of the following year. Certain things arose at the interview above mentioned at Derby, which necessitated my going to London. I had developed certain symptoms which had been diagnosed as "piles", and I decided while in London to visit a specialist in this class of complaint. On examining me, however, he discovered that I had an abdominal growth, and recommended me to see Mr Lockhart Mummery, the famous abdominal specialist who once went to India to operate on one of the Indian Princes, receiving as his fee the quite appreciable sum of £43,000. He confirmed the diagnosis of the first specialist and recommended that an operation should take place as soon as possible. Pursuant to this, I entered a nursing home in Manchester Square on the following Saturday, and on Monday of the following week I underwent an operation which was a preliminary and preparation for the really big one which was time for about a week later, i.e., in the first week of December, 1929.

I was told authoritatively that this was the most dangerous operation that a man was ever called upon to undergo, and that even in such skilled hands as those of Mr Mummery only 40% survived the operation. To give some idea of what took place, I may say that I was put under the anaesthetic at 10 am one day, and only recovered consciousness at 2am on the following day. Moreover, even with the 40% that survived the operation, the subsequent duration of life was exceedingly short, rarely extending to five years, and yet here I am writing about it in the year 1943, nearly fourteen years after the event.

The reason for this large percentage of deaths, either at the operation or soon afterwards, was due to the malignity or cancerous character of the growth, and Mr Mummery told me that for men of my age, 99% were malignant. However, it turned out and was confirmed by an eminent pathologist, that I was among the fortunate 1% where malignancy was not present.

Drastic interference with the human body such as this operation involved inevitably carries with it consequences of a troublesome nature, and in my case the usual consequence ensued in the form of bladder and prostate trouble. I consulted the principal surgeons of London in this particular department of surgery—Mr Jocelyn Swann and Sir John Thomson Walker—and they concurred in advising a further major operation of a kind similar to that performed on Mr Lloyd George, but weak as I was—too weak to be able to walk across a room—I did not act on their advice, and from what subsequently transpired I have no doubt I made a wise decision.

Mr Mummery did not charge me £40,000 as he did the Indian Prince, but he did charge me £345, in addition to which I had to pay the anaesthetist and several surgeons and doctors who, with more or less excuse, were in attendance at the operation. I paid Mr Jocelyn Swann a fee of one hundred guineas, though he did not operate. I paid about £100 to masseurs and at the rate of 26 guineas per week I paid the nursing home nearly four hundred pounds. I was sent to Champneys, the premier establishment of the Nature Cure cult, of which the Manager and High Priest was Mr Stanley Lief. I did this at the recommendation and under the supervision of a member of the ordinary medical profession. My bill there was £201 12s 0d, plus tips, and altogether from the time I left my home at Croesor late in November, 1929, until I returned to it in July, 1930, I had paid out over sixteen hundred pounds in connection with this operation.

I arranged with Mr Mummery to examine me every twelve months, and in December 1934 he again operated upon me at the London clinic. This operation being a sequel and a consequence of what had preceded it. On 14th of January 1938 I had a stoppage of the bowel which involved another major operation, this time at the Imperial Nursing Home, Cheltenham, and by Mr John Howell. During the course of this operation further mischief was disclosed, and instead of one, I really underwent two operations in close succession, but the strain of the two operations taken practically simultaneously nearly cost me my life, and the shock caused an increase in the blood pressure which in turn caused haemorrhage in the retinas of my eyes causing almost complete blindness.

This deprivation of sight I regard as one of the chief deprivations of my life, as it rendered it impossible for me to either read or write. I do, however, still sign my cheques and letter with a fairly legible signature, but this is done more or less instinctively without the aid of sight.

The next blow that I received at the hands of fate was on the 29th of January, 1940, when I had an exceptionally severe attack of shingles in the head, which also affected the upper part of my body. Both my doctor and nurse stated that they had never before experienced so severe an attack. I almost think that of all the illnesses I have encountered in my life, these shingles have caused me most pain and discomfort, and I am not free of severe nerve pain during any part of my waking hours, although more than two years have elapsed, and there is no indication of relief coming in the future.

The above is a pretty formidable and lengthy list of serious illnesses, but I believe that I have been stayed throughout by a strong will not to succumb, and a determination to carry on my business routine and a keen interest in the affairs of the world, including inventions, which, if exploited, I think have considerable value.

Even during my big operation in December, 1929, and the subsequent very trying months I personally sold the slates that were being manufactured at the Quarry, managed its finances, signed all cheques, and controlled its development.

I went a great deal to London for business interviews, and generally controlled my own personal affairs. My wife acted as my amanuensis and also accompanied me to all my business interviews, and I thankfully acknowledge the tremendous assistance she gave me.

I am blessed or cursed with an irrepressible sense of humour, and write nonsensical "poetry" even when I am suffering the most acute pain. (*See Appendix II*)

(Please note that the Editor, Cathy Woodhead, is not related to Moses Kellow's second wife. Ed)

36. NEAR ACCIDENTS

It is strange how one regards illnesses and serious operations as high danger-spots in one's life, and yet the chances of coming through, for instance, that major operation of December, 1929, were much greater than many a hazardous risk unconsciously taken or undergone during my life in the Slate Industry, where often the demarcation line between life and death has been but a hair's breadth. I will enumerate a few of these incidents.

When I was seven years of age, my parents lived at Tremadoc, and my father was superintending the driving of three trial tunnels at "Fridd-y-Bwlch", Blaenau Festiniog, which property adjoined the Welsh Slate Quarry on the east and the Llechwedd Quarry on the north.

He had sleeping and lodging accommodation on the Ffridd-y-Bwlch property and on one occasion—I was then seven years old—he took me with him there from Tremadoc.

One day he left me at the place where we lodged, saying that he was going into the trial tunnels and telling me to wait for him at our lodgings. He went up the mountainside to where the tunnels were being driven, leaving me behind, but I soon became bored with my own company, and decided to follow him. I did not know which of the three tunnels he had gone into, but decided to investigate the middle one.

This was rather a long tunnel, and became dark at a short distance from the entrance. I had no candle nor matches to give me light, but I found a short stick and by allowing the outer end of this to scrape along the side of the tunnel, I was able myself to keep in the middle. The tunnel was so long that I began to get frightened, but persevered to the end. On arrival at the end, I found a fully loaded rubbish wagon with two men already at the end of it preparing to push it out. Had I made my hazardous entry into the tunnel ten seconds later than I did, I would inevitably have been run over by the loaded wagon. As it was, all I got was a sound lecture from my father for my disobedience.

When engaged at the old Park Quarry, it was my father's practice to go home to Tremadoc twice a week, on Wednesdays and Saturdays, staying the other nights in a little wooden house just large enough to contain a bed, a table, and a cast-iron stove with the flue pipe going out through the roof.

On one occasion he took me with him, and I slept in the shack. In those days blasting powder was carried in open carts and not in closed, locked vans as today. A consignment of blasting powder arrived at the Quarry in damp condition, having been exposed to the rain in transit. The obvious thing to do was to dry it immediately, but there was no place to do this except near the stove in the shack, so two casks each containing 100lbs of blasting powder were brought inside, their tops removed to allow the damp to escape.

The stove, however, decided to take a more active part in the drying process, and shot out a spark from a log that was burning inside, sending it right on to the exposed powder in one of the casks. Fire at once began to spit all over the surface of the powder, and some of this jumped the gap and landed on the powder in the second barrel, and started the same process there.

My father coolly tilted the outer powder barrel a little on its side and rolled it out into the field. He then did the same with the other barrel, and when he had the two safely in the open, he calmly scraped off the burning layers of powder at the tops of each cask, and finally managed to put the fire out.

The utter recklessness of the whole proceeding was reprehensible. If that blasting powder had been just a little drier, there would have been two less members of the Kellow family living in the world. But my father knew, I expect, just how many chances he could take, and estimated the dampness of the powder to a nicety. He had great faith in himself.

I was sinking a shaft at the lower Park Quarry, which was 14ft square, and reached a depth of 5½ galleries. On one occasion I was standing at the bottom of this shaft with six miners, watching them at work. Twelve holes containing the highest explosive in use at the Quarries, ie blasting gelatine, had been charged into the bores and made ready for firing. The fuses were lighted and the miners and I got into a tub, to be wound up out of the way of the blast. But although the signal was given to start hauling, the engine did not function, and it looked for some time as if we were all to be blown to Kingdom Come.

Just when we were beginning to sweat, the driver succeeded in getting his engine started, but we were certainly not more than ten yards up from the bottom when the first of the holes went off, and most of the others had also detonated before we reached the top of the shaft.

In the same shaft, which I was sinking with the help of compressed-air drills, working at what was then the unusually high pressure of 90 lbs to the square inch, I was again with the men at the shaft bottom when suddenly the hose-pipe coupling, attached to one of the drills slipped its threads—probably due to careless cross-threading, which prevented it from being screwed home. Instantly, all the candles were blown out, leaving us in complete darkness. The freed hose-pipe began lashing about like a furious snake, the heavy brass coupling hitting the rock all over the place with sledge-hammer blows.

We all of us squeezed ourselves into whatever niches we could discover by groping in the dark, in order to try and evade the madly flying coupling, and as I crouched down in a tiny hollow I felt the shock of that brass coupling as it banged against the rock close to my head several times. Had it but touched my skull, it would have cracked it like an egg-shell.

No-one dared to move from his precarious shelter to try and reach the air-valve and shut off the pressure, and it looked as if that terrifying state of things was to continue indefinitely. At last, however, the air pressure ran down sufficiently to cause the hose-pipe to cease whipping about, and then, as if to add insult to injury, its open end fell into a puddle of dirty water that had collected at the bottom of the shaft, and blew it all over us. Speaking for myself, I had not a stitch of dry clothing on my body, and was caked in mud from head to foot. However, by what was almost a miracle, none of us had got hurt.

The shaft referred to was an exceedingly steep one, and it was intended to use it as the main artery of an extensive quarry. It had been planned to raise very heavy loads. This necessitated using wire ropes one and a half inches in diameter, and coupling chains made with links of three-quarter-inch diameter iron.

The end of the wire rope was fitted with a clasp securely fastened to the rope, but this clasp had to be attached to the coupling chain, and this could only be done by making a separate link, with an open end, passing it through the clamp, then through the end link in the coupling chain, afterwards welding the open ends together.

Obviously, this link required to be made as strong and safe as the coupling chain itself, and I would not trust the quarry smith to make the weld. I therefore sent to one of the foundries at Portmadoc, for a smith who had a high reputation as a welder. He came to the Quarry and so far as I or anyone else knew made a satisfactory job of the link.

The incline was put to work, and it was my regular practice to be drawn up in a wagon from a point four galleries from the top of the shaft. I had been drawn up one day, and had just reached my office when the inside foreman knocked on the door which I has just closed behind me. I re-opened it, and saw that his face was pale.

"Oh, Mr Kellow," he said, "how did you come up the shaft this morning?" "Why, I came up in the wagon as usual," I replied.

"Come inside the Quarry with me," he answered, and we returned to the shaft top, where I found that the link that has been supposed to be welded had never been welded at all, with the result that the link had gradually opened out until there was a gap in the end of it five-eighths of an inch wide, and the link of the coupling chain, which was made of three-quarter-inch diameter iron, was within one eighth of an inch of slipping through the gap. I vouch for the accuracy of this measurement, as I took it myself on the spot. Had the link pulled through, and the wagon run down that steep declivity, my life would have not been worth a farthing's purchase.

Once, at the Park Quarry, I walked in through the main tunnel into the No.1 chamber on the same floor. I found it in darkness and no men present, and wondered where they had all got to.

All at once, I found myself enveloped in a sheet of red flame, my ear drums felt as though shattered, and the breath was expelled from my lungs by terrific pressure. Stones whizzed past me at frightful velocity in all directions. I had been standing right in front of a blast. The men had retired into No.2 chamber, which was beyond the No.1, and had failed to post anyone to give warning to any person that might be coming along the main tunnel from outside. I carried away a vivid impression of what it must feel like to be near a bursting shell in an air-raid. But again I miraculously escaped injury.

On another occasion at Park Quarry I was removing some rock from a narrow face about one gallery above the main tunnel. Two men were working there, and one of them had just finished drilling and charging a hole. There was 4" of fuse sticking out of the hole.

The means adopted in Slate Quarries to give time to the men to get away when blasting is in progress, is to stick the fuse to the rock with a bit of clay, thereby covering the end where the train of powder through the fuse is exposed. A short end of candle is then placed under the side of the fuse, and the wrappings of the fuse have to be burned away before the flame can reach the train of powder in the centre of the fuse. When this occurs the fuse "blows", shooting out sparks, and it is then only a matter of seconds before the charge in the hole explodes. The workman who had charged the hole turned to speak to me, and in doing so inadvertently passed the flame of his candle under the projecting, and un-secured, end of the 4" fuse.

It "blew" instantly, giving us no time to get away, and there we were trapped, with no place to shelter. We had presence of mind enough not to attempt to escape in the line of shot, or we would inevitably have been hit. What we did was to fling ourselves flat against the very piece of rock that was being blasted, keeping of course as far away from the hole as we possibly could, but this was no more than 3ft In a few seconds the hole blew out past us, and neither of us was scratched.

Shortly after commencing operations at Croesor Quarry, I decided to take down some dangerous rock which formed the roof of the No.2 chamber on the "A" or main tunnel floor.

This chamber was approached from the shunting siding by an obliquely driven tunnel passing through the small vein which immediately overlies the older main vein, and not through the No.1 chamber, the floor of which had been worked away down to the "B" floor, vertically 45ft lower.

There was thus a large work-out space under the obliquely driven tunnel on the "A" floor.

One evening, I decided to stay up at the Quarry to superintend the operation of taking down the dangerous rock in the No.2 chamber. In the intervals of giving my instructions and inspecting the work I sat down on an explosive chest at the side of the obliquely driven tunnel, not suspecting that I was in any danger.

I left the Quarry at 9pm and the men who were working in the No.2 chamber told me that half an hour after I had left the floor of the obliquely driven tunnel and the powder chest on which I had been sitting, all dropped into the open space on the "B" floor, which was 45ft vertically lower down. Next morning I went to see the explosive chest on which I had sat the night before, and found it splintered into matchwood on the floor of the gallery below.

The galleries in a Slate Quarry are generally 45ft, and sometimes 60ft below one another.

The fore breasts in the chambers on these floors generally incline at an angle of 40 degrees, and a chain or chains are fastened to an iron peg let into a hole in the floor below, and hanging down over the sloping rock face along its whole length. This is to enable men to climb the rock face from one gallery to another.

One day I climbed up one of these chains and arriving at the top, I found the chain loose in my hands. Although it had been passed round the iron peg no attempt had been made to tie a knot in it, or otherwise secure it, but by pure accident the loose end of the chain was at the top, which was lying on the floor, just happened to be crossed by the long chain which was hanging down the rock face, and my weight upon it had pressed sufficiently upon the short end to prevent it from slipping round the peg, and thus precipitating me into the gallery below. When I reached the top my hands naturally lifted the chain from the ground, thus releasing the pressure on the loose end, which then slid away easily. The workman who left the chain in that condition was very much to blame, as anyone seeing a chain hanging down the rock face would naturally assume that it had been properly secured to the iron peg at the top.

At the Croesor Quarry there is an incline shaft rising about four galleries from the "A" or main tunnel floor, and at its top it connects with a vertical shaft 160 yards deep, which extends to the surface of the ground. At the junction of these two shafts there was the largest size of Robey semi-portable engine and boiler, mounted upon a bed-plate, the whole combination weighing about 12 tons. For convenience I will hereinafter call this the Robey combination.

The engine was connected to a pair of large-diametered winding drums and also to a pumping system through the medium of connecting rods and supports. Four galleries lower where the incline shaft reached the "A" floor another incline shaft leading to lower galleries had been sunk, and the Robey combination with drums and appurtenances were used to hoist the rubbish and blocks from the lower galleries up to the "A" floor level, where it could be sent out to the mills.

The situation of the winding gear made it very difficult to operate, as the driver could not see what was happening on the traffic incline, nor see the loads start or land, but had to depend on more or less vague signals to guide him on his operations.

I decided to bring the whole of the winding gear, including the Robey combination down to a point in the incline shaft a little above the "A" floor level, where the whole of the winding operations from the galleries below "A" floor would be visible to the engine driver. I planned to bring this 12 ton combination down the incline without dismantling it, and for this purpose I provided a large winch and a very strong chain. Flexible wire ropes were not then obtainable, and the links of the heavy chain took up so much room on the drum of the winch that less than 100ft could be wound upon it, and I wanted nearly 400 to reach the new position.

This meant fixing the winch no less than four times, and securing it to the rock so that it would not be pulled away by the weight it would have to control. I had given orders to the foreman to carry out the operation in the manner indicated, but on his own initiative he thought that he could save the time of one of the fixings of the winch by connecting four medium-sized chains to the end of the heavy chain I had provided.

I happened to arrive at the bottom of the incline shaft just as he was preparing to lower the Robey combination down the shaft. I walked up the incline and had arrived at the side of the Robey combination, but not the upper end, when I heard in succession four loud snapping noises. These were produced by the breaking of the four medium-sized chains one after the other, this being due to the fact that the lengthening chains were not each taking an equal strain, and so they broke one after the other as the full load developed on each in turn.

The result was that the twelve ton combination careered down the incline shaft that I had just previously walked up, and had I been but half a minute later in coming up, this combination would have swept me down with it, crushing me to pulp at the bottom.

As it happened, this accident, which would certainly have had fatal results had I been a few seconds later in arriving, helped us very much. Cinders and ashes from the boiler lay on the floor of the incline shaft all the way down, and the Robey combination soon collected a heap of these cinders in front of it as it swept downwards, with the result that shortly before it reached the bottom, it had a pile about 5ft high, which brought it to rest without doing much damage to the machinery. We only had to fix the winch twice instead of four times to get it in its final place.

The Croesor main level and mill floor are 1,563' above O.D., and I usually rode up to it on horseback. Occasionally, however, I went up to the Power House at the upper end of the Croesor Valley, and from there walked the short distance to the terminus of the Croesor Railway. The terminus was connected to the terminus of the Croesor Railway by a 700 yard long steep incline.

The motive power being derived from the weight of loaded slate wagons descending on the one side of the incline pulling a train of empty wagons on the other track alongside. The wire ropes were wound on a 9ft diameter drum at the top, and about a yard separated the full load going down from the empties and goods-wagons ascending on the other set of rails.

I decided one day to go to the Quarry along this incline route, and telephoned to the quarry for them to get a "crewling" ready for me, and to be in attendance at the top of the incline when I arrived at the foot. But by the time I got there, a dense mist had settled down, and I could only see a few yards up the incline, the rest was nothing but a grey blanket of cloud.

Sound signals intimated to the men at the top that I was ready to start up, and winding was started. The wagon in which I was riding was pulled up without mishap for the first 350yds, at which point it entered a cutting through rock. A stone on one of the rails in this cutting threw the wagon off the track and it ran against the rock at the side of the cutting, with the result that the whole crewling mechanism, including the wagons on both sets of rails, 6 tons of slate, 700 yards of wire rope and a very large drum at the top were instantly brought to a standstill. It is little short of a miracle that the wire ropes, coupling chains, hooks, and draw-bars, all stood the severe strain without breaking. Had any one of them done so, the wagon I was in would inevitably have careered down the steep incline to destruction, notwithstanding its derailment.

However, the men at the top realised that something unusual had occurred, and hurried down to investigate. The wagon was then put on the rail and the journey completed without further incident.

The large induction coil which had been presented to me by the Institution of Civil Engineers was capable of developing an electrical pressure of 750,000 volts. It was supplied through the house mains through a Caldwell Interruptor, by means of which the necessary pulsations of current were obtained and the voltage could be adjusted to any desired strength.

On one occasion I was using this coil for X-ray work at the highest voltage usable with the apparatus, some 400,000 volts, at which voltage the rays were exceedingly penetrative. I had the wires from the coil in my hands, and was making the contacts to the X-ray vacuum tube when my assistant, thinking that I had finished, switched on the current, and I got the full 400,000 volts through my body.

I was drawn up like a concertina, and I can assure you I was very badly shaken. My assistant, suddenly realising what had happened, and still having his hand on the switch, turned off the current in a fraction of a second, otherwise I would have been fatally electrocuted.

In working with these induction coils, although the voltage is very high, the considerable self-inductance causes impedance, which reduces the current (and it is the amount of current that passes through that matters), to low limits, and this fact, combined with my possessing a strong heart, certainly saved my life that day. I have had an experience that few others have had, and have survived to tell the tale.

One or two experiences still remain to be told, though they do not perhaps come quite within the title of hair-breadth escapes. In 1920, when in London, I received a telephone message from Major Gamwell, inviting me to accompany his two sisters and another young lady to Croydon for the following day—Sunday—where he proposed to make some aeroplane ascents.

37. FLYING AND SOME UNDERGROUND EXPERIENCES

Major Gamwell was the son of Mrs Gamwell, of Aberarto, Llanbedr, for whom I was acting as Engineer in connection with the Aberarto Hydro-Electric Scheme. He had been a flying officer in the Great War, and appeared to have access to, and full control of, a small aeroplane in one of the huge hangars. It may have been the aeroplane that he had piloted during the war, but I cannot vouch for that.

It had no military equipment, but was a two-seater machine and was described to me as being intended to carry a pilot in the back seat and an observer in the front. It was apparently a military machine adapted for reconnaissance work. The Major made several ascents, taking one passenger each time, and being the only male in the group of visitors, I naturally went up last.

I was astonished at the smallness and flimsiness of the machine. The body was constructed of light wood-work, which one would think ill adapted to stand the heavy strains it was called upon to bear, and this wooden framework was covered by a thin fabric. The pilot's seat, which was behind, as well as the observer's seat, had barely room for a man to sit down.

The pilot's legs rested on almost horizontal bars to the base of which were pedals attached, these being the rudder bars. Between the pilot's legs in a vertical position stood the "joy-stick," which was universally jointed at the base. I was given to understand that the rudder bars, operated by both feet, controlled the 'plane's direction, while the "joy-stick" gave the necessary tilt to the machine to enable it to turn round in harmony with the rudder control, as well as, by a forward or backward motion, causing the 'plane to rise or descend. A very simple but ingenious contrivance.

The most important instruments on those early machines were the altimeter and the revolution counter, showing how fast the aero-engine was turning at any given time. The art of flying apparently, consists in climbing an incline in the air to compensate for the pull of gravitation towards the earth.

If the nose of the machine is turned upwards too much it will fail to ascend the incline, and the 'plane is then said to stall. Stalling is exceedingly dangerous, as the machine will at once begin to slip backwards and downwards in the air, instead of forward, this being known as a tail spin, which is next to impossible to control or correct.

In this respect it resembles a man driving a horse attached to a trap. So long as the horse is going forward the driver can control his movements by pulling the rein to right or left, and stop it by pulling up both reins together. If the horse backs, however, it is useless to pull the reins, and if the trap is near a precipice and the horse is backing towards the edge, there is nothing the driver can do in all probability except jump out of the trap in time.

I was told by Major Gamwell that a constant watch has to be kept on the revolution counter, so as to be sure that the engine was turning fast enough to enable the machine to climb at any required angle.

I took the observer's seat with the Major's remark that should the revolutions drop below one thousand a minute the 'plane would inevitably stall, registered vividly on my brain. I found duplicate controls in the observer's cockpit, and these I assume were for the use of the observer—who would have to do his best with perhaps a limited knowledge of the art of flying—should the pilot be rendered hors de combat! His handicap seemed to me to be further enhanced by the fact that he had no instrument board to guide him as to what was happening to the engine.

In fact, what with the observer also having to keep his feet clear of the rudder bars and his knees from the joy-stick while the pilot was in charge, the whole machine rather struck me as a crude contraption ill adapted to carry human beings with much safety into high altitudes. Both the pilot and the observer had tiny glass windscreens in front of them, and not having an airman's helmet, I had my hat tied down on my head before I settled into the rear cockpit.

When flying at speed up aloft, if by any chance I moved my head the least bit beyond the small area protected by the windscreen, I felt I was not only in danger of losing my hat, but my head into the bargain.

A little boy once aptly described a hurricane as "the wind in a hurry," and it was soon brought home to me on that first aeroplane trip that the wind is in a very great hurry as it passes an aeroplane.

I cannot say that I experienced anything very unusual when climbing to an altitude exceeding 6,000ft, this being probably due to the gradualness of our ascent enabling my eardrums to adjust themselves to the lessened pressure unnoticed by myself. Quite the reverse, however, took place on our rapid descent. We came down in a series of spirals which rendered me at first somewhat giddy. The predominant sensation during this descent was a high pressure on the ear-drums which made me very deaf, so much so that when I arrived on the ground and tried to speak I felt like a man inside a barrel trying to talk to those outside. It took about two hours for this deafness to wear off.

From the relatively high level of 6,000ft I travelled over a considerable stretch of country, and looking downward on the houses below, they seemed like dolls' houses. A house seen from above appears much smaller than that same house seen in elevation from the gowned.

A few years after this first flight, I flew from Croydon to Paris in one of the large civil passenger planes then in operation. Neither the speed nor the height at which these planes flew was considerable according to modern standards, but nevertheless the journey was accomplished in a very short time. I was in Paris at 11am, which practically gave me another day for sightseeing.

On balance I do not think that the travel by air cost any more than the train and boat service, plus one day's accommodation. During the journey I found it pleasant to view the landscape below, and watch the great channel approach and then pass underneath us. The motion was smooth, and did not tire me in the least. Before descending, Wrigley's chewing gum was served out to each passenger, and he was advised to chew it, as the motion of the jaws helped to prevent the deafness which I had experienced on that descent back in 1920, with Major Gamwell.

A singular reminder, however, of the risks involved in air travel at that time was that the plane had doors on its top side through which passengers might climb in the event of the machine falling into the Channel.

Before the electrification of the Underground railways in London, I was at Bow Station during a pea-soup fog, and the trains were running very irregularly in consequence. I arrived at the station and had waited about twenty minutes for a train, when I could just distinguish through the murky atmosphere the lights of a train coming in on the opposite platform.

I knew it was the opposite platform because the lights were travelling from left to right. A few minutes afterwards I saw dimly the lights of another train entering the station in the opposite direction, i.e., moving from right to left. I naturally assumed that this was my train and that it was alongside my platform. The fog was so thick that the station was almost in complete darkness, and as I walked towards what I thought was my train I walked out over the edge of the platform and fell on the railway track below.

Had the electrification of the Underground been then in operation, I would have been electrocuted, as the third rail or "live" rail is mid-way between the two others, and a person falling so as to make contact between the live rail and the ordinary rail which serves as a return, would certainly have had a fatal amount of current passing through his body.

As it was, I hurt myself considerably. Other passengers on the platform helped me to get back upon it, I feeling very dazed and shaken, and suffering considerable pain. I was disgusted, and irritated almost beyond endurance, therefore, when a fanatical crank handed me a tract on the "Evils of Drink." All my friends know that I am a total abstainer, but I have no use for the jackasses who go round impudently thrusting themselves and their fat-headed opinions upon their fellow men, as this man did to me.

However, the sequel to this accident was no pleasant one. I returned to my hotel with great difficulty, and sent for a doctor. I had hurt my jaw rather badly, and the doctor said that I had ruptured some of the fibres of the biceps muscle in my left arm.

This had the effect of contracting the muscle so that my arm was drawn up around my neck, and the following night was one of agony. All business in London was, of course, at an end, and I decided to make my way home in two stages, breaking my journey at Chester.

I had to get assistance in undressing that night, and in dressing next morning, and the boots who assisted me in the morning said, very sympathetically; "It's quite all right, sir. I had another gentleman just like you last night." So he also diagnosed the cause of my trouble as "strong drink which biteth like an adder and stingeth like the serpent." I did not trouble to explain to the hotel servant that he was making a mistake.

At Portmadoc, on my return home I went to see my own doctor and told him what had happened.

"I believe I have broken some of my ribs," I said.

"Nonsense!" was his brusque answer. "You could not have walked into my consulting room as you did if you had broken your ribs, and besides that, the doctor in London would have been sure to have discovered it if you had." "Well—will you examine me then," I asked.

"I will do nothing so foolish," he replied.

Arriving back home, I tried to feel my own ribs, and besides feeling acute pain at a particular spot I thought I could actually feel the ends of broken ribs. I bore with this for a fortnight, and then went to Portmadoc again to see my doctor.

I met him in the street and accosted him directly with the remark that, say what he liked, my ribs were broken.

"All right," he gave in, "come in and I will have a look at you." This he did, and when he finished he exclaimed: "By jove—you have two ribs broken!" He then bandaged me up, and in due course I recovered. I had actually travelled from London, and walked about for two weeks afterwards, with broken ribs.

After this experience I am not likely to forget London pea-soup fogs, teetotal cranks, and my faith in the infallibility of doctors was considerably shaken. The incident of the tract distributor at Bow Station reminds me of a religious crank who once stopped me as I was walking hurriedly along Gracechurch Street, London, to ask me in sepulchral tones: "Are you saved?" I fail to understand why these things should occur to me, because I do not possess a red bulbous nose, nor, so far as I am aware, have I got a particularly sinful expression.

38. RESEARCH ACTIVITIES

To end this somewhat haphazard account of my life and activities, I would like to give some details of my research activities of late years, and give some of my ideas about improving the Slate Industry.

Early in this autobiography I stressed the advantages of 3-phase alternating current, especially for power work, and among these advantages was the robust and trouble-free character of the alternators.

This was due to the absence of commutators and brushes which are the source of nine-tenths of the troubles met with in operating electrical machinery. There was, however, one field of application in which it compared unfavourably with direct current because assuming a rail return was adopted for one of the phases, it still left two phases to be dealt with by overhead conductors, which created great difficulties at points and crossings. With the alternative method of direct current only one overhead conductor with a rail return was needed, and this was the system in common use for street tramways.

Though more convenient as regards the overhead conductors, the difficulty of commutation imposed a low limit to the voltage that could be used, 500/550 being the figures generally adopted, which meant very heavy transmission losses where long distances were involved.

The only feasible alternative was to use single phase alternating current, thus reducing the overhead conductors to one, as in direct current. Single-phase transmission, however, had the very serious defect that it involved the use of single-phase motors on the locomotive which had very little starting torque—obviously a crippling objection where trains had to be started from rest.

The ideal would obviously be a direct current generator which would give very high voltage without involving the use of a commutator and brushes. The discovery of how to get this result has been the dream of electrical men almost ever since Electricity has been in common use, but just how to do it has eluded them.

In a bi-polar dynamo there is a reversal of the direction of the electric current twice in every revolution, and in a multi-polar dynamo there is a reversal of the direction of the electric current twice in every revolution for every pair of poles employed. The use of the commutator is to make the current uni-directional as regards the outside circuit, and the problem was to eliminate the commutator and brushes and yet obtain the uni-directional current.

A particular design of direct current generator with commutators and brushes is in general use all over the world, and the fact of its universality has given rise to the general belief that any attempt to vary it by eliminating the commutator and brushes is fore-doomed to failure.

Earlier in this account of my life, I mentioned an amusing experience I had at Penrhyndeudraeth, where a flock of sheep attempted to pass my car, and, led by a particular ram, leapt five feet into the air over an empty space through which they could easily have walked, each sheep following their leader's absurd example.

Now "standard practice" is worshipped like a fetish by the majority of people, and like the sheep in the story, they regard any departure from it with disapproval and incredulity. I can recite the shibboleths regarding the impossibility and absurdity of being able to eliminate the commutator and brushes and yet be able to obtain uni-directional or direct current as glibly as the next man, but I have little respect for shibboleths as I know very often, just around the corner, is a fact either unknown or ignored which can knock all the shibboleths into a cocked hat.

I believe such a fact exists, and that it is possible on lines somewhat different to those prevailing in standard practice to design a dynamo conforming to the true laws of electromagnetic science, which will give a high electrical and mechanical efficiency without involving the use of the objectionable commutator and brushes.

During the years 1933-4 I prepared a design in conformity with this with full principal and detail drawings, and also worked out with great care the electrical, magnetic and mechanical calculations appertaining to it.

I patented this and then took it to the Chief Engineers and designers of most of the great electrical manufacturing companies of the Kingdom, and also to other experts. The first man I showed it to was the Lecturer on Electricity at one of the major English Universities, who was also the author of standard works on the construction of the DC dynamo.

He spent three hours with me in examining the drawings and calculations, and at the end of the period, said to me: "Well, Mr Kellow, if there is a 'nigger in the woodpile' I cannot find him." This, coming from such a man, encouraged me very much. Later on, I was agreeably surprised to get such a favourable reception from practically every important electrical engineer to whom I submitted my idea, that I felt sure I was on the right track of this elusive problem. The Metropolitan-Vickers Co, after having had it examined by their experts in London, and having expressed the view that my design was sound in principle, arranged with their works at Trafford Park, Manchester, to send their Lecturer to meet me in London to discuss the design in detail.

He arrived, and we had a two or three hour interview. I may say that at none of these interview did I meet adverse criticism or discouragement, but I was very much astonished to find that even among the greatest experts in the kingdom, and notwithstanding the vast extent of the electrical industry, that there was a divergence of views regarding even the fundamental principles involved in the electro-magnetic generation of current.

Under these conditions, and where an entirely novel design of dynamo was under consideration it was but natural to expect caution in expressing their views. The attitude I met with was certainly not hostile, but was that in a field of investigation so complex as that of electro-magnetic generation of current, no certainty could be obtained without making the machine and trying it out.

Well, that is as far as I got. In December 1934 I was laid low by my serious operation at the London clinic, and at the end of September 1935 I removed from North Wales to Cheltenham, the preparations for which, and the subsequent settling down naturally occupied most of my time.

My illness and preoccupations, and especially my almost complete loss of sight has prevented any further progress being made. Whether I have a design of dynamo in my plan cabinet worth a million pounds or more, I cannot say, but I would be quite willing to postpone the time when I shall be playing a golden harp and shining as a vocalist, if I could have a few years in which to develop this invention.

Between the years 1925 and 1935, I was working on the idea of producing synthetic slate from slate powder, particularly for electrical purposes, moulding it into the various sizes and forms required, during the process of manufacture. I made over 400 different experiments, the composition and treatment of each specimen produced being carefully recorded in books, so that they could be repeated if thought worth while.

The snag almost all through was that although produced under tremendous pressure I could not eliminate the voids, and the consequence was that the specimen was hydroscopic, or liable to take in moisture. The electrical resistance was quite satisfactory so long as the specimen was dry, but fell off to an inadmissible extent when it had absorbed moisture.

Success came in 1935 when I discovered how to produce a synthetic slate block which was hard, non-hydroscopic, had a beautiful, natural lustre requiring no enamelling or other covering to make it presentable and which moreover had the very astonishing property of possessing insulation resistance reading up to infinity.

Of course there is no limit to infinity, but taking it 100 megohms or 100 million ohms, it has an insulation resistance five hundred times that of the very best natural slate. Moreover, I could produce it at a cost that was only a small fraction of the cost of natural slate.

I took it to the General Electric Establishment at Kingsway, London, and there had an interview with the switchboard expert. I showed him the specimen I had, and he admired it very much. I asked him if he could test it for insulation resistance, and he replied in the affirmative.

He handed the specimen to his assistant and asked him to make the test in our presence. The assistant wetted the surface of the block, 2" diameter and 1" thick, with dilute sulphuric acid, and then covered each end of the block with tin foil, to which the leads of the "megger" were connected. He then looked at the reading and his eyebrows shot up. He turned towards us, and said: "It is infinity!" "No, it can't be!" exclaimed his principal. The two then looked at the instrument together, and the expert said: "It's infinity all right, Mr Kellow. You have something there that is going to be of great interest to the General Electric Co, Metro-Vick., and all the rest of us. Will you go down to Witton, Birmingham, to see our Managing Director, and if he is not there ask for the Head of the Research Department." As I was leaving the room, he said: "Mr Kellow, do you mind if I have that specimen again? I would like to test it once again." This was done, and he said: "Yes, it's infinity all right!" I went to Witton and saw the Head of the Research Department. My specimen was again tested and gave results similar to those obtained in London. He offered to place the resources of their works, including the use of presses that would give up to 7,000 tons per square inch for the making of larger blocks and slabs. Well, that is as far as I got with my synthetic slate, and the reasons for this have been fully explained already by my illnesses, other physical disabilities, and the impossibility of getting assistance in carrying out any further work.

However, I believe that in this case I might hand my discovery over to someone else to develop and exploit. It would be a pity to let all my work become of no use.

It is said that a lamb born on a particular field will never wander far away from it for the rest of its life, and this habit, known in Wales as the "cynhefin", has a considerable market value.

Now to some extent I am like such a lamb. I was bred and born in slate, and I have been connected with the quarrying and manufacture of slate almost throughout my entire life, and now towards the end of my autobiography the instinct to return to it induces me to express views which I have regarding possible future developments in the industry.

First and foremost I say that the waste in the quarrying and manufacture of slate is unnecessarily high. I am not far wrong if I say that for every 12 tons of slate quarried, only 1 ton goes into the market as manufactured slate, whereas I am convinced that by proper methods and careful supervision the proportion could be increased to one in six.

I cannot, of course, within the limits of this autobiography, go into precise details of the various means by which this could be accomplished, but a start could be made by opening up the working faces of the bargains in a different manner. Then when the rock face has been opened, the position of the holes from the open side should be very carefully measured and not dealt with in the casual and slipshod manner in which it is done now.

Moreover, regard should be had to the distance between the ribands (*Note: a riband is either a thinnish dyke running at an angle across the slate belt or a colour band*) as this will largely determine the length of the slate that can be made from it, and the width required will in turn be largely dependant on the length. Very often the rockman who plans the position of the holes has only a somewhat hazy idea of the sizes of slates that will be made in the mill from the rock he is taking out.

Again, some rockmen are more skilled that others, and where there is only one contractor in a bargain containing two rockmen, and he is the sole arbiter of how the bargain is to be worked, it is hardly to be expected that all these contractors have the knowledge, the calculating ability, and the carefulness always to do the right thing.

I once had a rockman in my employ who was generally regarded as very competent. He and his partner in the mill had a great predilection for making 20 x 10's and everything in the planning of the holes was directed towards producing this size as far as possible. The contractor is always what is known as the "bras-holltwr," or "big-splitter." If you observed him at work in his bargain, you would notice that nearly all his blocks had been pillared and trimmed to 1ft 2ins wide, and to ensure this the hole would usually be put at from 18 to 20 inches from the face side, or multiples of these measurements.

These blocks were very convenient for the slate-maker in the mill to handle, but after sawing them across, he would generally pillar a piece from each side of the block that was supposed to have lost some of its moisture since it had been quarried, and therefore less easy to split. That trimming and the subsequent dressing of the sides would bring the size of the manufactured slate down to 10 inches.

Now, all the piece of rock that had been blasted out was of as good slate as the 10 inches that went into the manufactured product, which was worth probably £10 per ton, and about 20 inches of the rock had been taken to produce the final 10 inches.

I wonder how many rockmen look on every piece that they trim off and assess its value in pounds, shillings and pence. Channelling, to prevent the rock being broken across by powder is so obvious a necessity that it should not require stressing, but what does require stressing is that it should be applied to the whole of the working face, and not utilized only to get a free bottom.

I will here make a slight digression. The rubbish wagons in general use are made with sheet iron bottoms, sides, and one end. The sides and end are rarely as much as 1ft 6ins deep, whereas the loads of rubbish are nearly double that. In order to keep the rubbish in, the sides and end have pieces of slate put up on end reaching to the top of the load.

A good number are required to go all the way round and it is very often the case that there are none available among the rubbish that is being loaded, except pieces of good slate-making quality. Hundreds of time I have seen a labourer take a hammer and chisel and split up a good block to get pieces to put around his wagon, and the Contractor looking on, or even helping him, actuated by the general idea that there is no alternative.

Then again the open end of the wagon has to be built up like a wall, and all this is to facilitate tipping when the load comes out of the quarry. As it is always the closed end of the wagon that comes down the incline, it is always this end that enters the chamber first, and the rubbish that is being loaded into it has to be lifted over the closed end.

These loads of rubbish when being tipped over, empty themselves very freely, and the proceeding seems superficially to be efficient. But I condemn the whole system. The sides and ends of the wagon should be high enough to hold all the rubbish the wagon has to contain, and tipping should be over the side of a very broad tip without necessitating the detachment of the locomotive from the train of wagons.

The commonest way in which the "big-splitter" (bras-holltwr) sends out his blocks from the quarry is in narrow pillars about 10 inches thick, the slate maker then divides this ten into two "fives", and the two "fives" into four "two-and-a-halves", each "two-and-a-half" being divisible, always in the middle, into sixteen slates. However, it is very rarely that these pillars are clean slate from end to end, and it will be found that even of the two-and-a-half thickness the riband starting from a point on the upper side reaches the lower side a foot further on, and the whole of this foot is cut out and thrown away although there is a large surface of clean rock on the upper side of the riband on one face, and the underside of the riband on the other face. This represents the minimum of loss, for often the ribands run obliquely along the face of the block, and anything from four to six inches in addition to the foot already mentioned has to be cut out and thrown away. Very often, where the cut is made opposite the riband on the underside of the block, the next riband on the upper side of the block is so close that a cut made in its line allows too short a distance to provide the length of a commercial slate, and a block, the great bulk of which is clean slate, is thrown away.

Common sense says that if there is clean slate enough between the ribands on the faces and backs of the blocks, this clean slate should be made available for manufacturing purposes, and it certainly can be done. Again, say there are three ribands running down through the block, with a greater distance between the first and second, than between the second and third.

Now obviously the pillared block, if wide enough to suit the longer distance between the ribands is too wide to suit the shorter distance between the ribands, and this is dealt with in the usual wasteful way of pillaring off all the sides and throwing them away—all this clean slate being worth maybe £10 per ton! Nearly all blocks when they come off the saw table are much wider than the slate that is eventually made from them, and this is trimmed and dressed off. Frequently blocks are sawn 5" instead of 2½", and when this is done it is necessary when cutting out a riband to cut out and throw away a piece of the block 2ft long instead of 1ft long, when the blocks are 2½" thick, and where the ribands run obliquely the loss is even greater.

When splitting a block, the thin splitting chisel is placed on its end as nearly to the middle as possible, and is then tapped on its end by a hard, boxwood mallet. This starts a slight split which is then continued across the end of the block by moving the chisel backwards and forwards and tapping it; and this is called marking the block. There is, however, nothing to indicate the exact position of the cleavage planes other than a general parallelism with the faces of the block as a whole, and the result that notwithstanding the best of judgement and skill has been exercised, the marking of the end of the block very rarely coincides exactly with the planes of cleavage.

The result is that following the marking from one side of the block to the other a split will be opened for, say, half to three-quarters of an inch in one plane of cleavage, and then break across into another plane of cleavage, leaving a little step on the surface of the slate. This is repeated until the other side of the block has been reached.

These little ridges extend downward from the edge of the slate for about two inches.

Meanwhile the short splits in different planes of cleavage emerging into one split in the true plane of cleavage, contributed to by the slight bending of the slate in the process of splitting.

These little ridges at the ends of the slate are called by the slate-makers "hoel cun," which means "chisel marks," but they are really more than that as already explained.

Now the significance of this is that during the subsequent dressing operation, if it is intended to make the cut very near the end of the slate, which is what economy demands, the slate will not lie evenly upon the supporting "travel," but will rest upon the ridges where the split jumped from one cleavage plane to another.

The result is that a somewhat jagged broken end is produced on the slate instead of a clean cut end. A slate is generally supposed to have a better and a worse end, and the end where the "hoel cun" ridges are is invariably the worse end, but it is not only a matter of appearance. The pressure of the dressing knife upon the unsupported slate between the ridges often starts little cracks in the slate, which can easily be detected if the end of the slate is wetted, and then allowed gradually to dry. Cracks like these, short though they are, are liable to extend under the influence of frost and eventually to cause a breakdown of the whole slate.

Some slate-makers, in order to avoid this, will dress of at least 2" of good slate in order to get rid of the "hoel cun". This will produce a slate with two clean-cut, nice-looking ends free from cracks, but at the cost of wasting 2" of the end of the slate. There are ways of reducing the waste from this cause, and I say that in the sub-division of the blocks as they come from the quarry, and in the splitting and final cutting of the slate to its market size, every process requires to be radically overhauled and changed.

This involves a change in the character of the manufacturing machinery, and of the methods employed. I have it very clearly in my mind what these changes should be and how they are to be brought about, but it is obviously outside the scope of an autobiography to attempt to describe them. Coupled with a change of methods there should be expert supervision by a well educated and intelligent foreman, and it should no be left to the often very casual and incompetent slate-maker to work his wasteful will upon material worth, as I have mentioned before, probably £10 per ton.

At the Croesor Quarry the mill men had contracts which were quite independent of the rockmen's contracts. Each contractor had a dresser whom he paid at day-work rates. This meant a separate contract for every two slate-makers in the mill. The blocks coming from the quarry were distributed with perfect fairness between the various contractors who, therefore, had an equal chance of producing slates, but certain slate-makers—the most competent and intelligent—would actually produce twice the value of slates for every ton of blocks that came into their hands of what the less intelligent would produce. I am aware that this is a striking statement to make, but it is literally true, and the comparison would hold good month after month for the same men. The incompetent men would in many cases not have a "bill" or value of output much inferior to those of the better men, but they would have used nearly twice as much tonnage of blocks to produce it.

It would have profited the owners better to have paid the competent men double wages than to have had the service of the inferior men for nothing, hence the need for expert supervision mentioned above. Nothing would give me greater satisfaction than the opportunity of renovating the quarrying and manufacturing of slate in the direction of efficiency and saving of waste, but how it is to be brought about now that I am 81 years of age is a problem not easy of solution.

Mr Winston Churchill advocates the increase of the population of this country above its 47 million level by increasing the number of births, but let us suppose that as an alternative to this that old men could be rejuvenated so as to have the energy combined with the accumulated knowledge and experience of age, the population would soon have a great leavening of supermen.

The difference between youth and age is simply that in youth the whole body is clean, whereas as life advances dirt accumulates inside the body which is not got rid of by the ordinary processes of elimination. All that is required to produce the ideal condition of the population as mentioned above, is simply to clean the dirt out and restore the body to the condition it was at birth. Is this unattainable? I state it as my considered opinion that within a comparatively short time, say 150 years, the span of life will be doubled. The extended period would not be one of decrepitude and senility, but of vigorous physical and mental health, teeming with the joy of life. The way in which this can be brought about is already within the knowledge of mankind. The deposition of waste matter, and toxins from the alimentary canal must be curtailed, and if these are introduced into the blood they must be efficiently eliminated through the ordinary channels of the lungs, the kidneys, and the skin, and a good deal of this elimination could be brought about by air-conditioning our houses, involving the cleansing, ozoning, humidifying and tempering of the air we breathe.

Now for extraneous assistance! What about a super-kidney? Taken, say, from an elephant, and through which the blood of the body could be passed and cleaned! I mention an elephant for choice because it has solved the problem of living up to 300 years, and so must have a pretty efficient scavenging apparatus, or kidney.

I dare say this will be regarded as very far-fetched and absurd, but the whimsical brain with which I am endowed prompts me to put forward "Jules Vernian" methods of apparently accomplishing the impossible, but nevertheless the idea of super-cleansing is a pointer in the right direction.

Most people have heard of the Voronoff so-called monkey gland method of rejuvenation.

Not long ago I had the opportunity of conversing on the matter with a responsible medical man, who had made a special study of the subject, and he told me that so far from being treated jocularly some truly marvellous results can be attained, but he said a recent discovery had profoundly altered the outlook for this treatment.

Associated with the glands were certain very potent chemicals, which could be introduced into the human body without introducing the gland itself, which after the exhaustion of its stock of chemicals became a useless and undesirable piece of foreign matter.

A renewed dose of the chemicals independently of the gland could be applied when the effect of the first dose became exhausted, and apparently perpetual youth could thus be secured. Probably here again the war has caused all experiments to be suspended and Hitler has one more count added to his indictment for his crimes against the world.

Now I would like to say to these would-be rejuvenators—"Please hurry up as much as you can, for I am 81 years old".

THE END

APPENDIX I
MAPS

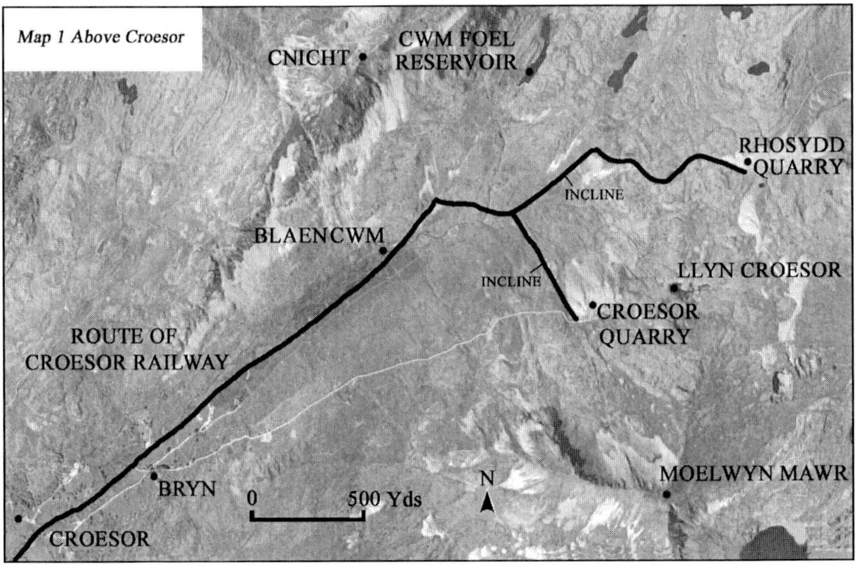

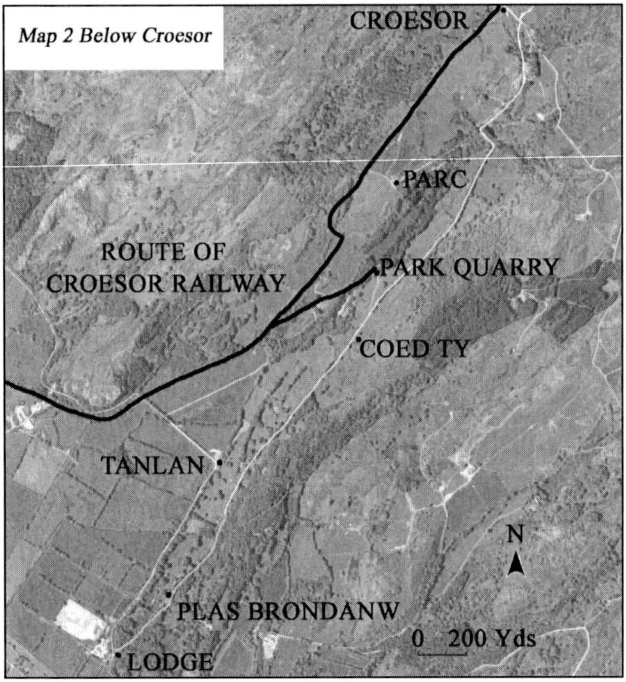

APPENDIX II

MOSES KELLOW'S BIRTHDAY POEMS

These two poems were written when Moses was staying at Penrhyndeudraeth. Clearly he preserved his sense of humour to the end, because shortly after his 81st birthday he returned to Cheltenham where he died only three months later.

The poems were published in the "Quarry Managers' Journal", with whose kind permission they are reproduced here. Copies are also in the General Collection of Mss. at Bangor, No.5276, p312 et seq. *Adrian Barrell*

TO COMMEMORATE THE EIGHTIETH BIRTHDAY ON 5TH JUNE, 1942, OF MOSES KELLOW

In the year eighteen hundred and sixty two
    On the 5th day of June mid a hullabaloo
A kid poked his nose in terrestrial life,
    This kind of thing happens when a man has a wife.
An onlooker gave him a jolly good whack,
    But the kid yelled defiantly I WILL NOT GO BACK,
I'm not paying a call I've come here to stay,
    And whatever you do, I WILL NOT GO AWAY,
I'm the offspring of William and Charity Kellow,
    A decent, respectable, well ordered fellow.
I claim the full rights of legitimate kiddy,
    So start the ball rolling by giving me tiddy.
Said William. I've got none, you'd better ask Cherry,
    She responded and soon he was happy and merry.
She said be sure you leave a small drop for the cat,
    Who I hear crying outside the door on the mat.
But he drank it all up to the very last drop,
    No power on earth could make that kid stop.
Now this kiddy at birth knew tonic sol-fa,
    For within the first hour he sang LAH LAH LAH.
Now if you ask ME if this RAYly was SOH,
    I'll answer in Welsh, DOH DOH DOH DOH DOH.
You may think this FAHfetched, but I'm still a TE, TE,
    And the answer is right as right can be.
The next thing to do was to find him a name,
    So they searched all through history for a man of great fame.
Among biblical Jews with long hooky noses
    They found a great leader with cognomen Moses.
In those ancient days they held the strong view,
    'Twas a passport to heaven to be named like a Jew;

And if the name came from a biblical source,
He just walked right in as a matter of course.
So the kid was named Moses with all good intentions,
Of Germans in those days they had no apprehension.
All this happened in Cornwall at Old Delabole,
Where you'll find a big quarry with a half mile wide hole.
Later on in his life there was turned a new page,
He was taken to Wales when but three years of age,
He there in due course attained man's estate,
Surrounded on all sides by mountains of slate.
This slate he dug out from deep holes underground,
And sold it to customers all the world round.
In THREE PHASE ELECTRIC he was pioneer,
And his name on numerous patterns appear.
The Inst. of C.E. have done him much honour,
Of which some may live when he is a goner.
For thirty-six years Merioneth he served,
And was put on the Bench, which was what he deserved.
Naughty boys were firmly punished that way,
Naughty men are treated the same way today.
Throughout he has spent a most strenuous life,
Had a talented son and a duplicate wife.
In nineteen two nine he met Mr Mummery,
Who told him straight out without any flummery,
That he had a growth that would settle his hash,
Unless it was cut out at a big cost in cash.
Now Moses at this time was aged sixty-seven,
And he thought it was too soon to go up to heaven.
So he let Mummery ply him with augers and saws,
And scissors and knives and what else no-one knows.
As a souvenir Mummery took out a chunk of his bowel,
And later a chunk was taken by Howell.
These men went to races and lost all their cash,
And then for poor Moses they made a quick dash.
They cut him in ribbons and bored him with holes,
If you saw his poor tummy you'd think they were moles.
Like his namesake of old this Moses was Holy,
And was teased by a Nurse with a big roly poly.
Abnormal enlargements at the base of the back,
Tempted sorely teased patients to give it a whack,
Or perhaps 'twas a pinch, 'twas very exciting,

But I cannot put down all the details in writing.
Nurses possessing this vulnerability,
Should learn to behave with decorous humility.
To return to our subject of fragmentary Moses,
His life was clearly not one bed of roses.
In this wicked world troubles don't come in singles,
He's haemorrhaged eyes and dreadful head shingles,
But you know what he said when he came the first day,
I'm not making a call but I've come here to stay.
So young ladies he takes for long country walks,
He won't allow loitering as onward he stalks,
He hasn't a heart so far as he knows,
And is supple and lissom from his head to his toes.
He's not lost much hair and he's not very grey,
And he hasn't a wrinkle that's what his friends say.
His nurse says he has a schoolgirl complexion, (1)
Not dependent on art to hide imperfection.
His memory is as good and his mind is as clear
As ever it was in his twentieth year.
He's an octogenarian in nineteen four two,
And though old he's as cocky as if he was new.

LINES COMPOSED ON THE EIGHTY-FIRST ANNIVERSARY OF THE BIRTHDAY OF MOSES KELLOW

Another year come and another year gone,
And Moses Kellow is eighty one.
In eighteen hundred and sixty two,
He began to paddle his own canoe.
Take eighteen six two, from nineteen four three,
The answer is eight one, that's as plain as can be.
In the past year his voyage has not been smooth sailing,
For Moses the skipper has been rather ailing.
A ferocious shark lurked down in a hole,
Intent on destroying him body and soul.
But keen hunters of sharks were on the alert,
And saved him from getting much bodily hurt.
He's now set his course towards nineteen five three,
But U-boats are about and tempestuous sea.
Editor Mac from motives fraternal,
Put his birthday poem in his popular journal.
Notoriety was thereby achieved,

I do not say fame for I'd not be believed.
But the thing that thrills me down to the core,
Is that canny Scot Mac is asking for more.
Every Q.M. has a thundering big head,
Into which tons of knowledge are constantly fed.
Now, although they're good men they'll be tempted to curse,
If instead is poured in my nonsensical verse.
At Cartrefle, Penrhyn, two years he has stayed,
And never experienced a bombing air-raid,
But in future I think that fair Penrhyn Town,
Will go up in the air and never come down.
Poor Helen'll be shot up so terribly high, (2)
She will make a permanent hole in the sky.
With momentum so great, she'll be shot up so far,
That for the rest of her life she must live on a star.
As Moses has travelled since he was a nipper,
He did not book seats in the fast astral clipper.
He went down instead to his Cheltenham home,
And I think for some time is not likely to roam.
When he walks outside on his exercise rounds,
He can still sense the beauty that lies in his grounds. (3)
There are fountains, waterfall, river and brook,
Allegorical figures in almost every nook,
Golden avenues, shrubberies, pergolas of roses,
All as designed by the gardener Moses.
Of course there's not the floral display,
Of pre-war Oakley, when in its hey-day,
Ten thousand plants, some of Royal descent,
Made beautiful views wherever you went.
Now you will find that food's grown instead,
For if you don't eat, you'll quickly be dead.
Potatoes and peas and all kinds of veges,
Are grown on the lawn and all round the edges.
This is all to the good and not to the worse,
As it helps to keep some cash in his purse.
Going back to his home of comfort and ease,
Every appointment is certain to please.
Lofty rooms with plenty of light,
Not only by day but also by night,
Fires electric and water that's hot,
Every convenience is found on the spot.

This is where Moses desires to stay,
    Until there arrives the Victory day.
Nincompoop meddlers are everywhere found,
    You can't move an inch, they cumber the ground,
But the war has been fought and the war has been won,
    So of course the social millenium's begun.
Some imbibed Beveridge, that made them feel gay,
    While others looked glum, for they had to pay.
They mustn't say no and they mustn't say nay,
    Or all that they have will be taken away.
If a person demurs, he'll be classed as a toff,
    His arms pinioned down and his head be cut off.
They'll quickly take one's skilled private nurse,
    Such swank will not be permitted of course .
They won't care a jot if the patient is worse,
    For they keep in reserve their communal hearse.
By a subtle trick of legerdemain,
    They'll turn debts into assets and spend them again.
There's nice air around but he dare not breathe it,
    It's Government air and requires a permit.
The kiddies at school, the poor little creatures,
    Have to bow and kow-tow to socialist teachers.
You may have a nice razor and want to do shaving,
    But blades are rationed and you'll be misbehaving.
You can't have a wash, for there's no controlled soap,
    Except a small piece at the Cape of Good Hope.
If you don't fill a form before kissing your wife,
    They'll put you in jail for the rest of your life.
The stupid restrictions will Moses so jar,
    That with Helen he'll go and live on a star.

Notes:

1. Moses' resident nurse was Miss Alice Mahon, a well-built Irish lady who was quite capable of carrying him up and downstairs single-handed. After his death she remained with Mrs Kellow and ultimately was well rewarded. See the section on the Kellow Wills (Appendix VII).

2. Moses was the eldest of seven children and Helen, his sister, was the youngest. Llys Helen, in Croesor, is thought to have been named after her. Helen married Dr. Hughes, of Penryhn, and is still remembered with affection. Moses' fear of her being blown up is a reference to the explosives factory at Penrhyn.

3. "Sense" rather than "see". By this time Kellow was blind.

Adrian Barrell.June 1995

APPENDIX III

APPLICATIONS TO ENGINEERING INSTITUTIONS

On 31st March, 1903, at the age of 41, Moses Kellow became a candidate for membership of the Institution of Civil Engineers. He was proposed by Gilbert Gilkes, and his application was supported by N.F.Wilson, R.C.Jackman, E.B.Ellington, J.Menzies and G.C.M.Hardingham.

Kellow states that he received his education and scientific training at the Davies Grammar School and by private tuition during the nine years from 1868 to 1887. In the following ten years he had been a pupil under and subsequently an assistant to his father, the late Mr William Kellow, a "General and Mining Engineer". There had followed sixteen years of engineering experience, as set out below; he was at present engaged as Engineer and Manager of the Park and Croesor Slate Quarries.

Work during pupilage and training: 1877-1887.

Mining work generally, including mine surveying; tunnelling; shaft sinking; bridges; winding engines; pumping plant; rock drills; and sawing, dressing and planing machinery—both in the office and in charge of practical work.

Subsequent Engineering experience: 1887-1903

The underground slate quarries, of which the Candidate is in principal charge, have been opened and equipped at a total capital outlay of about £250,000. Practically the whole of the present plant and buildings have been designed and carried out by him, and include steam and hydraulic winding engines; steam and hydraulic pumps; over-shot water wheels; Pelton wheels; three hydraulic pipe lines aggregating 6,500 feet; steam engines; air compressors; air winches; rock drills; five mills, fully equipped with sawing, planing and dressing machinery, and having a floor space of over 30,000 feet; and several miles of railway, including steel and other bridges. The whole of the mining work done for the last fifteen years has been to his plans and under his supervision; the mine surveying has also been done by him. The area of the mining properties under his charge is over 1,700 acres. He is at present engaged in laying out a three-phase electrical plant. As a first instalment, about 350 HP will be generated from a head of water of 860 feet, converted into electric power, transmitted one mile, and applied to winding, hauling, pumping, ventilating, lighting, and various other purposes. He has initiated and carried into successful operation an improved method of slate mining, regarding which he gave evidence before a Government Commission, a Report of which will be found in the Blue Books. He is a patentee of the following:- A saw for cutting slate; interlocking slate ridging; hydraulic rock drill; and a planing machine for slate and metal.

Moses Kellow was elected as an Associate Member of the Institution on 4th December, 1906. Five years later, on 29th March, 1911, he was recommended for Transfer to Full Membership of the Institution. At that time his present occupation was stated to be Chief Engineer of the Park and Croesor Slate Quarries, the Kellow Drill Syndicate, and Croesor Railway, etc. His supporters were J.S.Hughes, T.Jones, G.Gilkes, N.F.Wilson, J.W.Wyatt, L.J.Hunt, A.H.Preece, Llewellyn Preece, W.J.E.Binnie, W.O.Rooper and G.G.M.Hardingham.

Details of his education and training are identical to those given during his initial candidature. There follow two sections relating to his experience before and since Election to the Institution. The former is very similar to that set out above and below is the second section.

Engineering experience since Election. 1906-1911.

Mining, mechanical and electrical engineering work, connected with the mining for and the manufacture of slate, and the maintenance of the works, buildings and plant, at the Park and Croesor underground slate quarries. Development work at these quarries, consisting of shaft-sinking, tunnel-driving, and other extensions of a mining character; additions to plant, comprising an electrically-operated hydraulic pumping station for a pressure of 750 pounds per square inch; a system of piping throughout the quarries for transmitting hydraulic power to, and an installation of, Kellow Rock Drills and Channellers; sawing and dressing machinery, bridges, railways, etc. Including the above-mentioned extensions and additions, the total capacity outlay on these quarries approximates to £300,000 at the present time. The total area of the quarry properties is 1,797 acres.

Surveying, and the preparation of mine plans. The complete control of the Croesor Railway, involving the maintenance of the permanent way, rolling stock, and plant generally. The complete control of the Croesor Estate and the Park Estate, including erection of new buildings and plant, and maintenance generally.

The erection and complete equipment of the Keldril Engineering Works, which are railway connected, and electrically driven, lighted and heated. The most modern and efficient machinery has been installed throughout these works, including capstan, geared-head engine, and screw cutting lathes, planing, milling, grinding, drilling, and other machinery. Tool-room, store rooms, drawing and commercial offices, etc., have been provided. These works have an output capacity, at present, of about £800 to £1,000 per week of manufactured goods, but this is capable of substantial increase.

Chief Engineer of the Kellow Rock Drill Syndicate, Ltd., a Company which was formed in 1908 to manufacture and exploit his rock drills and channellers, for which patents have been secured in all the principal mining and manufacturing countries of the world, including Great Britain, U.S.America, and Germany. This position involves the complete control of the designs, of the manufacturing processes not only of the patented machines, but of other auxiliary machines and accessories which are manufactured in the Keldril Works and elsewhere. It also involves visits to mines, and advice to their owners, regarding the installation and maintenance of rock drilling plants.

The design of machinery for other Engineering Works, particularly in connection with his patent planer, saw tables, etc., etc. Consulting engineering work generally, including surveys of mines, inspection of electric and other plants, the making of reports, etc., in various parts of England, Scotland and Wales.

[The Candidate is the Author of a Paper published in Volume CLXX Min Proc Inst CE, entitled "The Application of Hydro-Electric Power to Slate Mining."]

Kellow was proposed for membership of the Institution of Mechanical Engineers by G Hardingham Esq on 13th June, 1913. His application was passed on 26th September and confirmed on 21st November. In his application Kellow describes himself as a mechanical, civil and mining engineer, and gives no less than three addresses:

1. For mechanical work: Keldril Works, Croesor.
2. For consulting and railway work: Bryn, Croesor.
3. For mining work: Park and Croesor Mines.

Positions Held

I have been Chief Engineer, and in principal charge, of the Park Mine from 1887 to the present date. I have been Chief Engineer, and in principal charge, of the Croesor Mine from 1895 to the present date.

The capital outlay on these mines to date approximates to £300,000. The area of the properties is 1797 acres. I have specially designed a substantial portion, and installed practically the whole of the existing plant and machinery in these mines, which includes steam boilers and engines, steam turbines, three-phase and direct current electrical machinery, high-pressure hydraulic pumps and turbines, hydraulic rock-drilling plant, air compressors, and manufacturing machinery occupying four mills having a total floor space of about 30,000 square feet.

I carried out the erection, and complete equipment, of the Keldril Engineering Works. I am Chief Engineer of the Kellow Rock Drill Syndicate, Limited, a Company formed to manufacture and exploit my Rock Drills, for which patents have been secured practically all over the world. I am the Patentee of Kellow's Patent High Speed Planer, and other inventions. I am in principal charge of the Croesor Railway, and primarily responsible for the maintenance of the permanent way, rolling stock, and plant generally. I practise as a Consulting Engineer, and make inspections and valuations of machinery. I designed and carried out the first instalment of a hydro-electric power system (intended eventually to develop 1,000 hp), regarding which I read a Paper, No. 3650, before the Institution of Civil Engineers in 1907, and was awarded a Telford Premium. I am a Member of the Institution of Civil Engineers and a Fellow of the Geological Society of London.

Much as I am a Moses Kellow enthusiast, I must admit that there were times when he was less than modest. A good example is the above description of the Keldril Engineering Works. It all sounds very different from the reality of the small original fitting shop at Croesor Quarry and, although the Works may have had a manufacturing capacity of £800 to £1,000 per week, the account books of the Kellow Rock Drill Syndicate tell a more realistic story. Adrian Barrell

The Kellow Rock Drill Workshop

APPENDIX IV
OBITUARIES

Moses Kellow died in 1943, at a time when the ICE had suspended the publication of obituaries as a war-time economy measure. However, obituaries were produced by the IME and the RGS; these appear later in this section.

Kellow's obituary in the Memoirs of the Institution adds nothing to our knowledge of him; in fact, the first part appears to have been written from his initial application form. However, it is reproduced here for the sake of completeness:

Moses Kellow, whose death in his 81st year occurred on 26th September, 1943, was a pioneer in the application of electrical power to slate quarrying and was patentee of the Kellow Rock Drill, the Kellow high-speed planer, and many other inventions. He had been a Member of the Institution since 1913 and was also a Member of the Institution of Civil Engineers, and a Fellow of the Geological Society of London. On the completion of his general education he received his early training in mechanical and mining engineering under the direction of his father.

He then became Manager of the Park Mine, North Wales, and twelve years later took over the managership of the Croesor Mine. He held these joint appointments until his retirement in 1935.

His activities included the formation of the Kellow Rock-Drill Syndicate, formed to manufacture and exploit his rock-drill, for which patents were secured practically all over the world. In addition he practised as a consulting engineer, his advice being extensively sought by numerous prominent concerns. He also devoted much of his time to research in connexion with the development of a high-speed dc generator, and only ill-health caused him to abandon his efforts. Another interest of his was the production of plastics, largely composed of slate dust, samples of which passed the severest mechanical and electrical tests. He also contributed papers to technical societies, dealing with his professional work.

My visit to the Royal Geological Society at Burlington House was even less productive. Despite the best efforts of the Archivist and Librarian all that could be found was Kellow's obituary which is, at least, rather more thoughtful than the one quoted above:

Moses Kellow, J.P., M.Inst.C.E., M.I.Mech.E., a Fellow of this Society since 1905, was distinguished as a pioneer of direct application of hydraulic along with local hydro-electric power to the mining and dressing of slate.

Brought up at Delabole in Cornwall, he went as a young man to Parc and Croesor Slate Quarries on the high Moelwyn plateau above Ffestiniog and during 40 years watched over their development into a successful undertaking. Living in the remote valley facing Cnicht he equipped the quarry workshop as a precision engineering laboratory, and as an inventor and adaptor of American and Continental quarry machinery he built up a wide practice as a consulting engineer. His quarrymen and fitters had been the local farm workers, and having brought them prosperity, he attained considerable influence as an adopted Welshman and became Chairman of the Merioneth County Council.

As a mine manager and civil engineer he made good use of geology in his profession and in a practical way made himself an authority on the extension and variation of quality of Ffestiniog veins of slate. He collected fossils with David Homfray* and was proud to act as cicerone to geologists who visited the Glaslyn valley to seek out localities made classic by Ramsay and Salter. During the last decade he lived in retirement, and he died at Cheltenham in October, 1943.

*David Homfray, in addition to his interest in geology, was a solicitor in Porthmadog and had held shares in the ill-fated Croesor New Slate Quarry Company.

Kellow's obituary in The Quarry Managers' Journal was published in the issue for October, 1943. There are several errors. Kellow was 81 at his death, and he died at Oakley, The Park, Cheltenham, not at Bryn. The reference to his lovely home was to Oakley, to which he returned in April, 1943, after a two years stay at Cartrefle, Penrhyn, where he dictated his autobiography which was subsequently published in the QMJ. I doubt if Kellow ever owned a slate quarry, and I have seen no other reference to his father as general manager at Delabole. There are several other minor errors, but at least this obituary was written by a friend. (Adrian Barrell)

Moses Kellow, J.P., M.Inst.C.E., M.I.Mech.E., F.G.S., M.I.Q., passed away at Bryn, Croesor, Penrhyndeudraeth, N.Wales, on September 26, 1943, aged 82 years.

On the anniversary of his birthday this year and last, Moses wrote "Lines" commemorating the event and sent them to us for publication in this Journal. The last appeared in our July, 1943, issue. At the Editor's request he had dictated his autobiography but it is doubtful now if we will have the pleasure of publishing it. He wrote to us just a week or two ago saying he was hoping to send it on soon. A few months ago he wrote saying that the man who was putting it on paper—we do not know whether he was supposed to type it or write it—would not number the pages and had got them hopelessly mixed up. It is a pity as it would have been most interesting.

In his 1943 "lines" (he would not have them honoured by the name of poetry), he says about his lovely home at Penrhyn etc, to which he had returned from Cheltenham where he had been for a few years:-

When he (Moses) walks outside on his exercise rounds, He can still sense the beauty that lies in his grounds. You see, poor old chap, he had become blind and could not see his lovely grounds, but could "sense the beauty".

And that was the reason he could not help his assistant to put the sheets of his autobiography together.

Mr Kellow was an engineer by profession, and became a slate quarry owner. He was a native of Cornwall, but went to Wales many years ago. His father was a mining engineer and general manager of Delabole Slate Quarries. That was some considerable time ago, of course, but it gives a reason for Moses, a born engineer, turning his activities to slate quarrying. But slate quarrying was just a means of livelihood; his mind was continually on the engineering side of the business and he took out patents for a number of inventions, amongst others: slate sawing table; slate planing machine; high pressure turbine; compound water turbines (master patent); high speed ball-bearings; epicyclic speed reducing gear; high pressure cocks; interlocking slate ridging; variable angle tile ridging; improved method of laying slates on roofs; and his favourite, the Kellow Rock Drill, which embodied a number of patents. His drill was hydraulically driven so its use was confined to quarries which had a big source of water carrying a heavy head. At Park Croesor he had that, but few quarries are so blest so it could never hope for any general use.

Forty years ago he installed a three-phase electric plant of 400 hp, the first in any slate mine or quarry in the country. Then came to Park Croesor a 30 bhp electric mine locomotive, the first used in Britain.

A "system" he was very proud of was one for leaving the quarry waste in the mine to support the roof instead of bringing it to be dumped on the surface.

Was he not an engineer rather than a quarryman? As far back as 1907 he was awarded a Telford Premium by The Institution of Civil Engineers for a paper on "The Application of Hydraulic-electric Power to Slate Mining".

Well, Moses has gone the way we must all go one day. I knew him for about 35 years—never saw him upset once; always smiling; always happy and full of enterprise. Always full of some new idea, generally engineering. Even in his later years when he was blind and more or less of an invalid, he was happy and contented; one to be envied.

S.McPherson.

Finally, two obituaries from the local newspapers:

From the Caernarvon and Denbigh Herald; 1st October, 1943.

Quarry Owner's Death.

The death has taken place at Cheltenham of Mr Moses Kellow, aged 82, the owner and manager of the Croesor and Parc slate quarries. He was a member of the Institute of Civil Engineers and an expert in water engineering. Mr Kellow was a prominent public man in Merionethshire, a justice of the peace and a past Chairman and Alderman of the County Council.

From the North Wales Chronicle; 1st October, 1943.

Mr Moses Kellow.

The death took place on Sunday at the age of 81 years of Mr Moses Kellow of Oakley, The Park, Cheltenham, previously of Bryn, Croesor, Penrhyndeudraeth.

Mr Moses Kellow came with his father from Cornwall at the age of three years to Wales, and after brief residences at Caernarvon and Tremadoc the family settled in Croesor, the father, the late Mr William Kellow, being manager of the Parc Quarry. The son took a keen interest in engineering and passed several important examinations. He succeeded his father as managing director of the Parc Quarry, to which was later amalgamated the Croesor Quarry.

A keen debater and public-spirited man, he became a member and later alderman of the Merionethshire County Council. His acumen as a financier resulted in his selection of the Finance Committee, a post he occupied for several years. He was a Manager of Croesor School in the days of the old school board, and he took a particular interest in the development and career of the pupils, many of whom have reached a high plane in several spheres. Of a genial disposition, Mr Kellow was popular with a wide circle of friends, and he used to love meeting and discussing current topics with his contemporaries. He was a fluent speaker in both languages, and often presided at various meetings.

Some eight years ago Mr Kellow went to live at Cheltenham, but when war broke out he returned to his native area and lived at Penrhyndeudraeth. He returned to Cheltenham a few months ago, then being in failing health. He is survived by his widow, three sisters (Mrs Williams, Mrs Bellis, and Mrs Dr Hughes, Plas-y-Graig, Penrhyndeudraeth), and one brother, Mr Josiah Kellow.

The only son lost his life on active service with the Air Force in the last war.

Adrian Barrell April 1997

APPENDIX V
THE KELLOW FAMILY

Moses Kellow was born at Old Delabole in Cornwall on 5th June 1862. His father, William Kellow, then aged about 29, was a foreman (some sources say manager) at Delabole slate quarry and his mother, Charity, was about 21; they were both born at nearby Tintagel. According to Moses' autobiography he was three years old when the family moved to Caernarfon and subsequently to Tremadoc, where they were living in 1871. However Moses' sister, Polly, told her granddaughter Cherrie, that she was four when the family moved to Wales, which would mean they moved in 1870 and not in 1865. Later they moved to Croesor, where they remained until 1892, when William became the Quarry Manager back in Cornwall at Delabole Quarry. By October 1895 William was seriously ill and he died on 24th October. Charity survived him and moved back to north Wales and she died here in July 1910.

In January 1892 Moses married Nell (Ellen) , the daughter of Thomas and Elizabeth Williams of Bryn, Croesor. Thomas Williams was the then manager of Croesor and at his death in 1894 or 1895 Moses took over the management of Croesor in addition to Parc; presumably it was at this time that Moses and Nell moved to Bryn.

Moses and Nell had one son, William born in 1892. He became a mining engineer but, after training as a pilot during the 1914-18 War, was shot down and killed on 16th September 1918 just two months before the end of the war. He was buried in Naves Communal Cemetery Extension. His mother died on 30th July 1919.

In December 1927 Moses Kellow re-married. He records that "the years 1926 and 1927 were prolific of business worries, and I began acutely to feel the need of the companionship and sympathy of a wife". His new wife was his house-keeper, Martha Fielding Woodhead, whose father was a landowner and mill-owner in the Huddersfield area.

In 1935 Moses and Martha left Bryn and went to live in retirement at Cheltenham at "Oakley", a very imposing house in The Park. It is still there. With the outbreak of war they were troubled by a fear of bombing and they returned to Penrhyndeudraeth, where they spent two years at Cartrefle, also known as the Old Post Office, in the High Street and it was during this time that Moses dictated his Autobiography. The Kellows returned to Cheltenham where Moses died on 26th September 1943. His wife survived him until she died on 25th August 1951; in his will she was provided with an income for life, after which the money reverted to the main stream of the Kellow family (see Appendix VII).

Moses Kellow had two brothers and four sisters. Moses was the oldest. Mary-Anne (Polly) was born in 1865 and she married John Williams (see more about Polly and John below). Josiah Kellow was born in 1871 and he lived at Tan Y Bryn in Croesor; in 1891 he was a pupil-teacher. He later became a coal merchant in Porthmadog, being the principal shareholder of Kellow & Co, Coal Merchants of Portmadoc. During WWI his horses were taken away for war work and he went bankrupt. He appears in the Parc and Croesor story and at one time he seems to have been the Company's representative in Porthmadog, in charge of the Croesor Siding. The correspondence suggests that the two brothers were not always on the best of terms. Josiah had one son called Walter, a bank manager, who had no children.

Moses' second sister was Emily Maria Kellow born in 1874. She married Mr Bellis and they lived in Bromborough, Cheshire; they had no children. John Kellow was born in 1877; he became a minister with the United Free Methodist Church and died in 1900 of

TB in South Africa. The third sister was Helen, born in 1881. She married Dr Hughes of Penrhyndeudraeth in 1907 after Helen had contracted TB and Dr Hughes saved her life by collapsing her lung. In 1905 Charity, Moses' widowed mother, was living at Minffordd with Emily and Helen. Finally there was Elizabeth Jane who became a school teacher possibly at Tal Sarnau but she died aged 22, probably also of TB. Years later in 1943 Polly went to live with Helen and Polly died in 1946.

Of the six of Moses' siblings only his sister Mary-Anne (Polly) who had married John Williams had children whose offspring are still going. They had five; Ifor who died when he was a baby, John Arthur was born in 1899, Cherrie was born in 1903, William Trefor Williams was born in 1906 and Alun was born in 1907. We know most about William Trefor. He was always known as Trefor and after Moses' only son was killed at the end of WWI Moses took Trefor under his wing and Trefor helped in the quarry and later when it was shut down. Trefor had six children: Cherrie born in 1932, Jean born in 1934, John born in 1936, Gwenfair born in 1939, Emily born in 1947 and Hugh born in 1951.

Another Kellow is Joseph, the "Jos Kelley" mentioned in the 1873 Caernarvon and Denbigh Herald article as the manager of Parc. Presumably he was the same Joseph Kellow who wrote "The Slate Trade of North Wales", published in 1868, in which he is described as a quarry engineer of 26 years practical experience. Joseph Kellow was born on 24 May 1813 in Trevena, Tintagel. Moses mentions a visit to his cousin Martin Kellow, whose father was Joseph, however the records do not show that Joseph was Moses' uncle.

Kellow family group taken outside Helen's house, Plas y Graig, Penrhyn
L to R : Moses Kellow, Martha Kellow (second wife) Helen Hughes (Kellow's sister, married to Dr Hughes of Penrhyndeudraeth) and Dr. Gwennie (Dr Hughes' junior partner)
(Nanw Owen Collection)

APPENDIX VI

THE KELLOW GRAVE

The grave is in Tan-lan cemetery, Llanfrothen (SH620430). Starting from the tall granite pillar above Bob Owen Croesor's grave, the Kellow grave is two rows further in and six rows lower down. I am grateful to Miss Gwen Rowlands for helping me to find it. Surprisingly, the headstone is of marble rather than slate. Presumably Moses wanted the very best for his wife and only son.

The first two and original inscriptions are:

In loving memory
of
Ellen (Nell)
the wife of Moses Kellow
Bryn, Croesor
who died 30 July, 1919,
aged 51 years.
Also of
Lt. William Kellow, R.A.F.
son of the above
who died serving King and Country in France
on 16 September, 1918.

When Moses died his body was brought here for burial from Cheltenham. The inscription, in slightly different lettering, reads:

Also of the above
Moses Kellow
died September 26th, 1943
Aged 81 years.

Martha was cremated in Cheltenham, but her ashes were brought here to be interred in the same grave. The final inscription, in totally different lettering, reads:

Also
Martha Fielding Kellow
died August 25th, 1951
Aged 74 years.

So they lie there all together, looking up at the tips of Chwarel Parc, and beyond to Bryn and Chwarel Croesor

May they rest in peace.

Adrian Barrell January 1997

APPENDIX VII

THE KELLOW WILLS by Adrian Barrell

In the course of my search for Kellow's missing research papers I obtained copies of the relevant Wills. Three are available. Moses Kellow's Will of 6th July, 1928, made at the time of his marriage to Martha, is in the Gwynedd Archives (XD8/4/1596); this Will was subsequently revoked. His last Will is dated 24th August, 1943, just one month before his death on 26th September. Martha Kellow's last Will is dated 18th May, 1949, with three subsequent codicils on 15th December, 1949, 10th July, 1950, and 16th July, 1951. Martha died on 25th August, 1951.

The latter two Wills are held at the Probate Office.

Some of the provisions of the Wills are set out below, insofar as they are relevant to the Croesor story or provide details of the Kellow family. I have tried to balance the free availability of these Wills against a respect for the privacy of the Kellow family and other surviving beneficiaries; in most cases their names have been recorded but I have deliberately omitted the amount of the bequest.

MOSES KELLOW'S WILL OF 6TH JULY 1928

This Will was drawn up after Kellow's marriage to Martha. The Executors and Trustees were Martha Fielding Kellow, Josiah Gayer (Kellow's cousin) and Adrian Stokes, a Porthmadog solicitor. Presumably it replaced an earlier Will in favour of his first wife, Ellen, who had died in 1919; this earlier document has not survived. In 1928 Kellow was resident at Bryn and was employed as the Manager and Engineer of the Parc and Croesor Slate Quarries; Parc had closed in 1916, but closure at Croesor was still eighteen months in the future. In this Will Kellow made adequate provision for Martha, but only during her lifetime. She was to receive £1,000 per annum, which was to be raised from capital by the Trustees. She was also to receive "all furniture, books, scientific and mathematical instruments, chattels and all other personal effects, motor cars".

Other bequests were to his niece, Edyth Williams, his private secretary, Alfred W. Davies, and to the quarry clerk, Robert Owen. The Trustees were bequeathed £500 "to be applied by them in the provision and maintenance of a Recreation Room for the benefit of the inhabitants of Croesor but without imposing upon my Trustees any liability or obligation for the carrying on or maintenance of such recreation room and my Trustees after the erection of such building may hand over the same to a properly constituted and responsible committee elected by the inhabitants of Croesor over the age of twenty-one years to provide and carry on such Recreation Room at Croesor".

Kellow's remaining estate was to be divided into 14 equal parts. Helen Hughes, his sister, was to receive five parts. Another three parts were to be used by the Trustees to provide a lifetime income for another sister, Mrs Mary Ann Williams; after her death the capital was to be divided equally between her two sons, Trefor and Alun. Another sister, Mrs Emily Maria Bellis, was to receive three parts. Two parts were to go to cousin Josiah Gayer, and the remaining part to nephew Walter Kellow.

On Martha's death the residual estate was also to be divided into fourteen parts and distributed as above.

MOSES KELLOW'S WILL OF 24TH AUGUST 1943

This Will is a much more professional document than the 1928 version; the Midland Bank was appointed as the Executor and Trustee. When the Will was drawn up Kellow was resident at Oakley, The Park, Cheltenham. He and Martha had moved there from Bryn in 1935; they subsequently returned to Penrhyndeudraeth for two years, during which time the Autobiography was written, and they made their final return to Oakley in the April of 1943.

In this Will Martha received £5000 absolutely. This was in the form of shares in the following companies: W.Hancock, Ltd., Marks and Spencer, Siemens, I.C.I., Indestructible Paint Co., Imperial Tobacco Co., J. Summers & Sons, Shell Transport and Trading Co., and Spiers and Pond. Any shortfall was to be made up from London Electricity Supply Co. shares. Kellow expressed the wish that these investments should not be altered without proper professional advice.

Small bequests were made to Bob Owen of AelBryn, Croesor, to Edyth Williams (Kellow's sister), to John Evans, Tyn-y-Celyn, Llanfrothen, to Nanw Owen (daughter of Bob Owen), to John Griffiths, Bryn Siriol, Penrhyndeudraeth, (to whom Kellow dictated the Autobiography), to Nurse Ada Williams, High Street, Penrhyndeudraeth, and to Mrs Maggie Owen, Port Meirion, Penrhyndeudraeth.

The Calvinistic Methodist Church at Croesor received £20, but there is no mention of the earlier provision for a Recreation Room in the village.

William Trefor Williams (Kellow's nephew) was bequeathed "my scientific surveying and mathematical drawing instruments my collection of fossils rock specimens microscope and microscopic slides my experimental apparatus and synthetic slate block my technical note books and technical printed books to the extent sanctioned and approved by my wife".

Nurse Alice Mahon, who was Kellow's resident nurse both at Cheltenham and during his stay at Penrhyn received £200, and Kellow expressed the wish that "she should remain with my wife for the rest of my wife's life performing the same duties she at present performs for both of us.

With the arrangements already made she will find no reason to regret her association with the Kellow family". Alice was indeed well rewarded in the provisions of Martha's Will. £20 was given to Mrs Sarah Dodson, who was Alice's sister and sometimes stood in for her. After her marriage she moved to Rugby.

Oakley was left to Martha on trust but "with regard to the use and enjoyment of my said dwelling house garden and grounds and the receipt of the income of my residuary estate I wish it to be clearly understood that my intention is that such use and enjoyment and receipt of the income shall be for my wife alone and not under any circumstances for other persons". Similarly, Martha was to have life use of "all my household furniture motor cars plate plated articles linen china books pictures statuary and works of art garden tools and effects".

After Martha's death the Trustees "shall stand possessed of my residuary estate as to both capital and income thereof upon trust to divide the same into ten equal shares. Two parts were to go to Helen Hughes, Kellow's sister, two parts each to William Trefor Williams and Dafyd Alun Williams (brothers, and Kellow's nephews), one part to Walter Kellow, and one part to Josiah Gayer (cousin). The remaining two parts were to go to another sister, Emily Maria Bellis, but in this instance Kellow adds that "as it is my express desire that this should be for her personal benefit and as a provision for her old age and possible infirmity provided she shall give to the Company within three calendar months of my death a written undertaking that she will not give either directly or by Will any portion of her said shares bequeathed hereunder to the Bible Students Movement".

In the final paragraph of the Will, Kellow places on record "my great indebtedness both for my life and comfort to my dear wife Martha Fielding Kellow. I cannot express in words my sense of her devotion to my welfare. My object in this Will is to ensure as far as possible that she shall be able to live at Oakley and to maintain a position there equal to that we now enjoy together and I have therefore considered and wish her interests to be considered by the Company as of paramount importance".

A note after the signature states that the Will was signed "in the presence of us both being present at the same time who at his request in his presence and in the presence of each other have hereunto subscribed our names as witnesses the said Will owing to his blindness having first been carefully read over to him when he appeared fully to comprehend and to approve the contents thereof" The witnesses were Ian D. Yeatman, Solicitor, Cheltenham, and W.A.Wintle, Clerk to Rickerby Mellersh & Co., Solicitors, Cheltenham.

Moses Kellow died on 26th September, 1943, and Probate of this Will was granted at Gloucester on 31st January, 1944.

MARTHA KELLOW'S WILL OF 18TH MAY 1949

At some time between 1943 and 1949 Martha had left Oakley. Perhaps the house had proved far too large for her "alone and not under any circumstances for other persons", although she would have maintained a domestic staff. Whatever the reason, Martha moved to a smaller house, Tara, 100, Leckhampton Road, Cheltenham. Presumably the agreement of the Trustees was necessary. She was resident at Tara when her Will was drawn up.

It seems either that Martha had no immediate relatives or that there had been some sort of family rift, because there are no bequests to the Woodhead side of the family. Indeed, the first and by far the largest bequest is to Nurse Alice Mahon who received £4,000; certainly Alice found "no reason to regret her association with the Kellow family". Sarah, Alice's sister, received £100.

Lesser but still substantial bequests were made to "my husband's nephew, Arthur Williams", to "Grace Williams the wife of my husband's nephew William Trefor Williams", to "Mary Williams the wife of my husband's nephew Dafyd Alun Wyn Williams", to "Walter Kellow my husband's nephew" and to "Howard Gayer the son of Josiah Gayer".

Small bequests were made to Maggie Owen of Port Meirion, to the Revd. Hanson of Cheltenham and to "my dear friend Fred Hales 39 Tilehouse Lane Knowle in the County of Warwick"; Marjorie Ellen Hales, Fred's wife, received a painted chest from the dining room at Tara. "My household linen to Nurse Alice Mahon and my late husband's two nieces Grace Williams and Mary Williams". Moses Kellow's gold watch was willed to Walter Kellow or, in the event of his death, to Arthur Williams, but see below.

All Martha's residuary estate was to pass to Helen Hughes, provided that the amount did not exceed £1,000. In the latter event any excess was to be shared equally between William Trefor Williams, Davis Alun Wyn Williams and Walter Kellow.

The Will was witnessed by T. Bateman and D. Miles, both of 2, Promenade, Cheltenham, and both described as Gentlemen.

The Will has three Codicils. The first (15.12.49) records a change of detail and the second (10.7.50) adds a small legacy to Dr. Clifford Sadler of Cheltenham.

The third Codicil (16.7.51) reduces the legacy to Arthur Williams, but the main provision is a substantial increase in the legacy to Alice Mahon. Not only is her main legacy increased to £4,250 but, if the residuary estate exceeds £1,000, Alice is to receive an additional £250 before the three-way division described above. "These increases are given by way of an acknowledgment of the devoted care which she has extended to me in my recent illness". For the same reason, Dr. Sadler is now to receive £200.

Norman Rosier of the Midland Bank Executor and Trustee Co. is to receive £10 "as a small acknowledgement of the letters which he has written to me on my late husband's affairs from time to time and which I am afraid I have so often failed to reply to". Finally, "I revoke the gift by my said Will of my late husband's gold watch and chain to Walter Kellow and in place thereof I give the said watch and chain to Alun Wyn Williams my thought being that he will be able to pass this gold watch and chain on to his son David in due course it being my wish that he should do so".

Martha died a month after this final Codicil. Probate of her Will and Codicils was granted at Gloucester on 7th December, 1951.

Adrian Barrell January 1997

Moses and Martha in the ground of Oakley. Late 1930s
(Nanw Owen Collection)

APPENDIX VIII

EXCERPTS FROM THE ORAL ARCHIVES

Extracts from a conversation on 17th February, 1996, with Mrs Sali Kidd of Llanfrothen. During the 1920s Mrs Kidd's father was "Johnny Run", in charge of the horse-drawn trains on the Croesor Tramway.

I remember Miss Woodhead, his (Kellow's) second wife, before they got married; she was his housekeeper. I remember they used to come for walks round. She used to come down the railway from Brynhyfryd ... she had the route ... nearly every day; she was fond of walking, wasn't she? She had big earrings ... they used to jingle and we thought, "What a smart lady".

Adrian Barrell December 1997

Extracts from two conversations, in September, 1995, and February, 1996, with Mr Iorwerth Roberts of Llan Ffestiniog. Mr Roberts knew Croesor Quarry before the Second World War, and worked there with Trefor Williams from 1946 to 1948.

I knew Croesor from childhood, from my early teens. Trefor Williams married a cousin of mine. I was extremely interested in things mechanical, even as a child. Trefor knew all the answers! He used to describe the workings of all kinds of machinery, which I found fascinating in those days. Then I was in the army and Trefor suggested that when I was demobbed I should apply for a job as an electrician. I got it and I was there for about three years, from 1946 to 1948. It was a useful transition from army life back into civilian.

I remember Kellow's locomotive but I never saw it running. There is a little story attached to that. Trefor Williams drove the loco past the terminus (A6E) and let the pantograph shoot up. It was taken to the shop and repaired and he did exactly the same thing again! What was it John Williams said Kellow said? "A man who makes a mistake is a man." said Kellow to Trefor, "A man who makes the same mistake a second time is a fool".

Adrian Barrell January 1998

The following recollections have been extracted from some conversations which I had with the late Mr John Henry Williams of Penrhyndeudraeth during 1993 and 1994. Mr Williams worked at Croesor from 1921 until production ceased in 1930.

Yes, he was a great gentleman. As it happened, I was sort of his "pet boy". I remember quite well. Kellow was a character. He was very keen to get whatever slate you could get out of a block. I was working the third table from the entrance into the mill and he had visitors; he had a lot of visitors, he should have taken over Llechwedd! Just to show you, I had a big slab of slate on the table, and it had been sawn. Now then, this slab wouldn't make two 24 x 12s; the measurement was 23" so it was too small to make two 24s because you wanted to divide it into two and dress it again. He came in and asked me what I was going to do. "Well, I'm going to pillar it here", I said. So I put the 12" mark on there, put the chisel on and split it. Right, now then, this one was making 20 x 10. Kellow asked me, "Why didn't you split that in the middle and make two 22 x 11s? You've sawn that 24" long and now you're going to cut this down to 20 x 10, so you've lost four inches". Well, I considered that. I wasn't going to give in. It's our practise to make 24 x 12 and 20 x 10; they were the most commonly made slates. Anyway, I calculated; I was having a shilling for a 24 x 12 and sixpence for the other, that would be eighteen pence. The other way would have only given me seventeen pence. Kellow said, "You're right there, but you've still lost four inches in that length". That was the type he was.

Yes, Kellow was a genius. I think he invented a carpet sweeper. He wanted samples of slate to go somewhere. We were dressing the slate, and you catch hold of it like that. I remember him saying, "You can't get that off. Look at the mark of the slate; you can't get it out". So what he used to do was to dress the slates with gloves on, then he would take them down to Bryn and put this vacuum cleaner on them. There wasn't a speck left. It was powered from Blaencwm.

They were hand-churning, weren't they, in the old days, to make the butter? And Kellow said, "Well, I'll put in a Pelton wheel for you". So he made a small lake up above Fron Wen, and brought a pipe down. And of course he was down there experimenting, and he fixed it to the churn. And it started, and the churn busted. That's perfectly true; too much speed, you see, shattered the bucket.

When Kellow started to generate electricity at Blaencwm he offered it to the village for £5 a house. Of course, everybody was afraid; it was lightning, wasn't it, more or less? They refused at the time. I remember my father telling me. £5 was the cost, but how much they would have had to pay for it after, that's another thing, isn't it? But, for putting it in, he offered it for £5.

Kellow was a very interesting gentleman. He was my hero. I've nothing to say to Bob Owen and his fellow-scholars—it's the literary aspect that seems to count today. But I'm sure that the people of Croesor should have done something. Kellow put the organ in the chapel; that organ is still there. He was responsible for that. They should have done something years ago, when he died; they should have done something for him. It's a great pity, because Kellow has been responsible for the living of many people in Croesor and the district all around. He supported them for years; he gave them work. It was small, of course, but then that was the trend at the time. He used always to say, "If you can make money, then make it. If you are working hard and you make a lot of slates, you get a lot of money". That was his motto. Whatever you do you can get, but if you don't do it you can't get it. That was his theme all through.

Adrian Barrell January 1996

The late Mr John Watkin Jones, who was born on 15th July 1904 and worked at Croesor Quarry from about 1920 until the closure. Mr Jones kindly provided these reminiscences of his time at Croesor.

I worked for over ten years with my father, John Samuel Jones, all that time in the D16 chamber. Of the 62 men working in the Quarry as per the Time Book at the date of closure in December, 1930, (closure came completely without warning) only David James Williams and myself are alive today. I will be 93 years of age on 15th July, 1997. I witnessed many changes during those ten years. I came to know Mr and Mrs Kellow when quite young. They used to park their pony and trap, before their motor-car days, at my grandparents at Coed Ty, while Mr Kellow would visit the Parc Quarry, which was working in my early years. John Williams, from Penrhyndeudraeth, was the chief clerk there, helped by the famous historian Bob Owen, of Morfa Glas, before he became married and moved to live at Ael-y-Bryn, Croesor.

During my ten years working at the Croesor Quarry, seldom did we have the company of Mr Kellow there and, when he did attend, he always seemed more concerned with the performance of the marvellous drilling and channelling machines of his own invention—the Keldril and the Kellow Channeller. Although the machines gave such good performance, he wanted deeper holes bored in less time. This caused friction between Mr Kellow and his mechanics Zachariah Morgan and his son, John Emrys Morgan. One problem was that the pressure seemed to bend the boring rod, making it difficult to withdraw it from the bored hole. Yet Mr Kellow could be very human and friendly with us, especially so when we showed keen interest in his creations. He would explain and demonstrate to us. To quote an instance: calling a group of us to his foundry one day to demonstrate how electricity worked, he was pleased with our concentration. There were four or five Johns in that group. One of us, over-anxious to prove his interest, cut across the speaker with an unfortunate problem. Said he, "We are told, Mr Kellow, that if you rub your hands vigorously back and fore on the back of a cat, you will produce electric waves". With a smile came the response, "To tell you the truth, John, experimenting with cats is not in my line!" This reply was much to the amusement of our crowd, and to the embarrassment of John Jones.

Whilst working at the Croesor Quarry I had varied experiences whilst working with the Keldril boring machine and the Kellow Channel machine. This was mostly in the lowest depths of the mine at the D16 and D18 chambers. The boring performance of both the types of machines was unique, as recorded in The Croesor File. Mr Moses Kellow's engineering achievements are always extolled.

As a layman, I would comment upon some disadvantages we experienced whilst operating the machines. The Keldril is not a light piece of mechanism, especially when it had to be hauled up a surface of rock strata to a seat of performance, of ten or several metres height. In general we had to ensure assistance to haul it up with chain or rope to the actual seat of operation. The operator would have ensured that two parallel holes had been bored, by the hammer and gimlet method, about seven inches deep. Into these two holes the two bolts attached to the Keldril fitted perfectly; it was then stabilized with the thrust rod attached for screwing against the rock surface.

Since the vibration of the machine in its operation could be severe, especially when the cutter at the tip of the boring rod came across a vein of sulphur in the strata, it was of vital importance that minute preparation was made. This preparatory work often occupied much time and patience. The length of hose-pipe from the main water line to the aperture on the Keldril needed minute attention by the assistant. If the attachment was not securely screwed on, then when the main valve was opened the hose-pipe would break loose drenching in the operator and his assistant unmercifully, ending in quite unhappy controversy especially if the boring had to be done quite high up on the rock surface.

Other unpleasant events occurred when the operator was being pressed, especially by Mr Kellow himself, to exceed the times attained at boring different depths of holes. Often the water pressure of the rods, tipped with hardened cutters, would cause slight bending of the rods, especially at 6ft or 7ft depths. Then the extraction of the rods became a problem, resulting in the loss of working time. Sometimes even the cutter at the tip of the rod would break loose at the bottom of the bored hole. Working with the Kellow Channeller was a much calmer operation. Standing on firm ground, or on a planked stage if the channel had to be done on different heights of the rock face, one or two could move the machine step by step along the field of operation. Here time had to be taken to scrape the bottom of the "cut" of all slate debris from the cutters. Channelling operations saved the quarry from the loss of good slate material. The prime advantage from the Keldril and the Kellow Channeller was the complete absence of slate-dust, the enemy of the slate-quarrymen. Mr Kellow's own views on this are quoted in The Croesor File.

Living at Bryn, the Kellows were very much part of the community. Martha Kellow showed great sympathy towards anyone involved in an accident at the quarry. Moses Kellow was a very good geologist. Once you showed some interest in the subject there was no end to him—he would explain all the different layers of the slate area.

When our daughter Rhianwen was born Martha Kellow brought a little kitten for her. We always called it the Kellow kitten.

I asked about the block of granite in C9 which Kellow used for Keldril experiments.

That's right, yes. Because he wanted to develop his Keldril so that it would bore through very hard rock, like granite. Oh yes, he had some big ideas about that drill. Mind you, I think that killed the quarry, investing so much money in the Keldril. That was the general opinion at the time. There had been rumours that the quarry wouldn't go for long for some years before it closed.

Finally I asked about the day of the closure.

Kellow appears to have been reluctant to face the staff, so he sent Bob Owen in his place. Bob Owen, yes. I remember the day quite well. It was all so sentimental-like. You would think there'd be a funeral! I remember one of the chaps from Tremadoc saying, "We're not having a funeral!" We had no wages or anything, it was a hard ending. The North Wales Quarrymens Union, they tried, but we could get nothing, the Company was bankrupt. We had nothing.

Bob Owen's account of the same incident, kindly translated by his son, Tudur Owen, goes as follows:

The worst job I ever had was having to tell the men that there was no money in the Company's vaults to pay them. The Company had become bankrupt and, instead of going to give the news to the men himself, Kellow sent me. It was approaching Christmas. I had no idea what to say. I said, "Now boys, keep hold on your spirits. There's no pay for you today". I had expected abuse but what came was pure silence. I remember the men walking quietly and slowly down the mountainside, rather than running as they used to do.

Adrian Barrell December 1997

The following pages are the result of several conversations which I had with Mr Edgar Parry Williams, of Croesor, during June and September, 1997, both at his house and during walks in Cwm Croesor. Mr Williams was born in 1932 and has lived in Croesor all his life; both his father and his grand-father worked in Croesor Quarry. He is the nephew of the late John Henry Williams who kindly provided me with so much information on the quarry. He spent much of his working life as a shepherd on various farms in the area and has an encyclopaedic knowledge of Cwm Croesor. Latterly he studied at the University College of North Wales at Bangor; then, employed by Meirionnydd District Council but based at Coleg Harlech, he spent two years on an oral history project for the St Fagan's Museum, recording information from older agricultural workers. Time has now overtaken him and he himself is being recorded; he is only four years older than me so it will be my turn soon!

On one of our walks we were accompanied by Mr Iorwerth Ellis Williams, of Llanfrothen; born in 1926, he was the last turbine driver at Blaencwm and was able to fill in many details. I am very grateful to both gentlemen for giving me so much of their time and their knowledge.

The people of Croesor had a deep respect for Kellow, who was regarded as quite a genius; they had great faith in anything he devised. Kellow's father-in-law, Thomas Williams, was a very religious man and a great benefactor of the area. He was involved in the early development of the quarry and of Croesor village; it was he who built the Chapel. He would have been held in high regard. Kellow stepped into his shoes and was similarly respected. Kellow's first wife Ellen, Thomas Williams' daughter, was also very much part of the community in which she had grown up; she helped at the school and with sickness or any other family problems in the village.

One of Kellow's offers mis-fired. The generating plant at Blaencwm was not used at night or at weekends, so Kellow offered a free supply of electricity to the Chapel which, in addition to Sunday services, was the focal point for all events in the village. The matter was put before the elders but, after due consideration, the offer was refused with thanks as they felt that electricity was too much like bringing lightning into the building. Kellow did bring electricity to Bryn and to Moelwyn Bank. Bryn was a show-place. Edgar lived there at one time and recalls that the old electrical fittings were still in place, including a large stone slab on which the fuse-boxes had been mounted. Bryn is remembered as a cold house, very expensive to heat and run.

In Bryn there is one room on the ground floor which faces the quarry road, giving a good view of who was going up to the quarry; the quarrymen left their bicycles leaning against the nearby wall and it was possible to see who was going to be late for work. In this office there was a little cupboard built into the wall, with a lockable door; it looked like a cupboard for the storage of beverages but this was not the case, for Kellow was teetotal. Potential customers for slate would be brought up by car to Bryn; Croesor slate was rather inferior to the Ffestiniog Old Vein product, so it was necessary to put on something of a show to impress them. Early in the morning of the day when a customer was due Kellow would send a message up to the quarry specifying what slates he wanted. John Henry Williams, Edgar's uncle, would then make the slates, so that they would have a full fresh sheen. These slates would be wrapped in a special square of silk and put in a special leather holdall bag to protect them. Then they would be brought down to Bryn in good time and put into the small cupboard. At the right moment they would be brought out again and shown to the potential customer, probably with the suggestion that they had been there for years.

Edgar was able to enlarge upon his uncle's story of Kellow's turbine at Croesor Fawr. The turbine was installed in a small building between the dairy and a barn. Not only did it drive the churn in the dairy, with the disastrous results already recorded in The Croesor File, it also drove a chaff cutter in the barn. It is said that Kellow attempted to speed up the feed to the chaff cutter by thrusting a broom into the hopper, only to be rewarded with a pile of very short pieces of broom handle.

Kellow's status in the village was such that he was automatically asked to chair any special events. On 8th May, 1926 he was President at a concert in the Chapel. A poem, in the strict Welsh metre, was written in his honour by Carneddod; I am grateful to Edgar for translating it:

I MR MOSES KELLOW, LLYWYDD YR HWYR

Enwog fedrus beiriannydd, - o dalent
Deilwng fel dyfeisydd;
A champau ei riniau rydd
Glod i'w enw drwy'r gwledydd.
Da gwr, o ddysg diguro, - mynwesol,
Gymwynasydd effro;
A'i fraint yw noddi ei fro, -
Coeliwch - "eich cefn" yw Kellow!

TO MOSES KELLOW

A very famous engineer,
Of exceedingly good talent as an inventor.
His expertise in this field is well known in all the nations.
A good man of excellent education;
His joy is to benefit his neighbourhood.
Remember, our support comes from Kellow.

Edgar pointed out the ruin of the barn which had been the "cowshed barracks" mentioned by Tegfelyn (a pseudonym of a quarryman who wrote a series of articles in 1862). The external dimensions of the barn are 10' x 21'. Edgar explained that in those days the cut hay was not carried very far; he pointed out no less than five now-disused barns along the valley floor. The Tegfelyn barn was the furthest up the valley; it was the one most useful to the quarry and the one the farmer would be most likely to let them use. Internal fittings were not taken away in anticipation of further farm use. Farmers were here both before and after the quarry! When both Rhosydd and Croesor were working many Rhosydd men came through the roofing shaft down into Croesor because the Croesor smith was much better than his opposite number at Rhosydd. When a wedding was planned the quarrymen would collect for a present; they would pay the smith to make them a fender, stool, poker and tongs, everything for a hearth. Steel was expensive, and the Croesor smith was well known for making good use of old scythe blades; when they had worn too thin for their original purpose he re-fashioned them into knives, while old cut-throat razors were made into blades for pocket-knives. He was also an expert maker of three-pronged spears for salmon poaching. Moses Kellow and Ifan Jones, the Manager of Rhosydd, were both Magistrates, and they very often had brought before them poachers who had been using implements made in their own quarry workshops and smithies.

Kellow's economy carburettor, which injected water-vapour into the mixture, was inspired by the well-known phenomenon that cars seem to run better on a misty day; Kellow is said to have noticed this while driving from Bryn up to the Quarry. When Kellow went to Cheltenham in 1935 his Morris Oxford was left at Cwt Injan, just below Croesor village. The local children would put it into reverse gear and use the starting handle to wind it uphill from the barn, subsequently joy-riding back down the hill into the barn where the wall served in lieu of brakes. At the beginning of the Second World War Trefor Williams went to Cwt Injan to see if the carburettor was still in place, thinking that it might be of interest to the Government. The carburettor had gone, and a rumour spread that it had been stolen by German spies. Certainly the local CID questioned local car owners, including John Henry Williams.

David Brindley, Kellow's chauffeur, lived at Brynhyfryd; he married Edgar's mother's cousin. In pre-motoring days Kellow had a coachman; he too married one of Edgar's mother's cousins.

Trefor Williams is remembered as the builder of an experimental windmill. This was erected on a step-ladder by the river in Croesor. All the children from the school came out to see the demonstration but the steps had not been anchored properly and at the first gust of wind the whole device blew over. Towards the top of Cwm Croesor, near the Bwlch Rhosydd path, there is a spot called "The Hill of the Windmill". It is just possible that this may have been the site of a Kellow experiment in wind-power.

Adrian Barrell November 1997

Excerpts from a lengthy interview with Mr Jack Morgan (aged 77) on 9th August 1978 held at Snowdonia National Park Study Centre, Plas Tan y Bwlch and included here by kind permission of the interviewer, Michael Lewis. Jack Morgan went to Croesor Quarry in 1919.

On Moses Kellow: He was one of those men who would spend other people's money like water and not his own. It was the same with the first car, an old Morris Oxford, Bullnose, two-seater, with the hood on. It was in the garage for years, he wouldn't sell it. He couldn't get enough for it until it rusted and was no good to anybody. The same thing happened to the steam engine in the old fitting shop, he wouldn't sell it to the scrap merchants, he wanted more, in the end he got nothing, somebody else got it, but he got nothing. That is the type of man he was. He was very keen. Very tight with his own money.

He was a crafty old chap. He had the Kellow Rock Drill Syndicate account with the ironmonger in Penrhyn, the Croesor Estate Account and the Croesor Quarry Account, whenever he wanted something from there, which account do I put this on? "Oh," he said, "put on the Croesor Estate Accounts." It was the estate that paid for everything. He was a real devil. The Syndicate never would pay anything, they didn't have any money to pay.

There was one good thing about him, he had the patience of Job, he was a good tutor.

On an improved planing machine: You see he had to set his plane, it was a very fast plane and you had to set it when it was planing, you had to stop with the reverse to go back again, but it was coming back faster than it was planing. Well the mill was there, the plane was there, the railway here and here the river, and old Bob Jones was working the plane and old Kellow had been with him and told him, "Be careful, I want you to give a good demonstration for these people to see how it works." Old man Kellow came with these people. "Are you ready, Bob?" "Yes, sir." And he jammed the lever, then rrrrr... and the running table was in the river.

On domestic electricity: When he had the old car, he always ran the battery down, so we had to take the battery up to Bryn to charge it. Now we had to make something to reduce the current down, so we made a board and had various lamps on it to get the right current for charging, and the only way to connect the wires onto the battery was to bind them with a copper wire. And I happened to be doing this, and I had to turn it to get it tight, and one end touched the other and it burnt my finger, it went red hot, and I showed it to Kellow, and you know, he said, "you silly fool, you shouldn't do that". One of the lamps had gone off, and he went meddling, tried to get it off, the ring to screw the part of the lamp on, he got that loose somehow, still had the current on. He said, "you put that back, Jack," and I was shaking … I managed to do it, but he wouldn't try it. You do it all the time, that's his way, he was always on the safe side, don't you worry.

Notes on a conversation on 29th September 1995 with Mrs Nanw Williams of Criccieth. Mrs Williams is the daughter of Bob Owen Croesor; in 1938, when she was 14, she went to Cheltenham to work as a maid to Moses and Martha Kellow.

I don't remember much, you know; it was nearly sixty years ago. I went there on the day I was 14 and I was a maid there for three years; there were four maids altogether. I got only five shillings a week. It was a very nice house. He had a big shop in Cheltenham in to furnish it—he used to go and see everything before he bought it. It was a show-house for the Cavendish House shop.

It had a lovely garden, too. There were three or four gardeners there. At the bottom of the garden he had a pump, a fountain, with the switches in the house. When he had visitors there he made me stand by the switch. Then he used to say, "When I sing Hen Gwlad, the fountain will go on, but when I sing God Save the King it will stop". He used to enjoy that. Yes, that was the party trick.

Nanw Williams (L) née Owen and Alice Mahon (R),
Moses Kellow's full time nurse (Nanw Owen Collection)

Then the Germans were starting to bomb down in Cheltenham and the other girls were leaving, leaving me on my own ... they were Welsh girls. I remember telling them that I wanted to leave too, and he asked me to go to the office with him. He was telling me why I shouldn't go, and he ended by giving me half-a-crown and saying, "Thank-you for not going". And I told the other girl who was leaving; she let my mother and father know, and they said, "Come home if you want to". I don't know why, but I was angry with this girl for not minding her own business, and I stayed until they came up here. It must have been a change for him to come to Penrhyn from that lovely house, and the gardens. I used to go in the office with him when he was doing the wages. He had a great sense of humour. I remember him laughing; "What are you doing?", he said. And there I was looking at a penny to see how to spell "George"!

I used to have to read to him from one of my father's books, and translate as well. He was nearly blind and couldn't hear very well. I don't think he knew much Welsh, and as he couldn't hear I had a problem there!

Adrian Barrell January 1998

Oakley (55 The Park, Cheltenham) as it is today

Notes on a conversation on 20th July 2015 with Mrs Cherrie Carlisle of Caswell Bay, Swansea. Cherrie is the great-niece of Moses Kellow, being the granddaughter of Moses' sister Polly

I remember going to Bryn where there were oceans of white carpet and I was made to take my socks off and say I was sorry. I had queried why they needed to have so much of this white carpet and was told off for rudeness.

I also remember going to Moses' mansion in Barmouth when I was about 5 or 6 years old. There was a lunch party and Nanw was serving. I remember sitting on Moses' lap and then being taken to the shops by Martha and the nurse where they bought me a new beret. My sister Jean went to Barmouth on a different occasion and she was also bought a hat.

From an email on 26th June 2015 from Nia Povey regarding memories of her father Gwyn Roberts.

Gwyn Roberts' father ran the garage at Garreg. Gwyn has a childhood memory of Kellow's American Marmon car calling at the garage to fill up with petrol - this would have been in about 1931. Gwyn well remembers serving petrol to Brindley, the chauffeur, with Moses sitting aloof in the front of the Marmon. Moses never spoke to Gwyn who was about twelve at the time as all contact was made through Brindley. The car was black or very dark blue and very grand.

Cathy Woodhead July 2015

A 1931 Marmon Straight Eight car similar to the one Moses had. Marmon Motor Car Company was an American automobile manufacturer founded by Howard Marmon and owned by Nordyke Marmon & Company of Indianapolis, Indiana, US. It was established in 1902.

APPENDIX IX

MOSES KELLOW by Rodney Weaver

This article first appeared in the Festiniog Railway Society Heritage Group Newsletter No.7, Summer, 1986, and we are grateful to the Group for allowing us to include the article here.

One of the most remarkable characters in the history of slate mining around Blaenau Ffestiniog was Moses Kellow (1862-1943), Manager and Engineer of the Park and Croesor Quarries from 1895 until their final closure in 1930. Kellow had in fact taken over Croesor before 1895—a quarry with a quite fascinating history of stop-start operation mixed with some quite remarkable engineering achievements—and it was he who amalgamated the two operations in the Croesor Valley.

Kellow inherited a traditional slate mine at Croesor and proceeded to modernise it. His two major achievements were electrification and the perfection of what is still the most powerful portable rock drill ever made. Croesor's mill was originally powered by a 39 ft waterwheel located in a huge wheelpit beneath the floor of the mill, and one of Kellow's first acts was to replace this by a Pelton wheel. He then turned his attention to the use of electricity and quickly came to the conclusion that the only distribution system worth considering was three-phase alternating current (ac) at a fairly high voltage. This was diametrically opposed to contemporary thinking in Britain, where vested interests clung to the low voltage direct current (dc) system promoted by Edison.

Sebastian de Ferranti was the only major champion of high voltage ac, but he believed in single-phase distribution. Kellow, however, looked east to Hungary where the firms of Ganz and Koller were perfecting three-phase technology.

Equally revolutionary was Kellow's belief in the use of three-phase induction motors to drive his machinery. Many believed this to be impractical and the usual technique was to convert to low-voltage dc by means of a motor-generator and use dc series-wound motors. Kellow persevered, however, and effectively designed his own electrical system for Croesor. A small power station was built at the foot of the inclines down to the Croesor Tramway, from which power was supplied not only to the quarry on the hillside above but to Kellow's house at Croesor Village—a building which is, I understand, a story all to itself. Three-phase motors were installed to drive the machinery and, as Kellow had predicted, proved wholly satisfactory. Not content with this, he equipped the main adit level of the mine with overhead electric traction and built his own electric motor to work it—the first electric locomotive in Wales and one of the first in Britain. Sadly, by this time events had overtaken Kellow; when he designed his plant in 1901 he was a lone voice, but by 1904 opinion had swung in favour of three-phase distribution and the standards adopted at Croesor were different from those eventually adopted nationally. One would like to know if it was Kellow who in 1906 persuaded Bruce Peebles & Co to defy convention and offer directly-coupled three-phase haulage motors to Oakeley Quarries, for the installation there was one of the earliest to employ such things.

More fascinating still was the Kellow hydraulic drill. Pneumatic drills had been introduced in some mines by 1895 but Kellow did not like them. They were slow, uncomfortable to use and created much dust, and he recognised the health hazard this caused. Kellow belonged to the minority school of thought that the more efficient way of drilling holes in rock is to use a bit rotating at low speed under a high axial load and lubricated with water, and he proceeded to build a drill based on this principle that would be light enough for one man, or a man and a boy, to carry round a slate mine. His first, made in 1896, used a small Pelton wheel supplied with water at a pressure of more than 500 lb/sq in. This was concentric with the drill shaft and drove it by means of epicyclic gears, the shaft itself being loaded by a piston in a cylinder on the back of the drill and fed with water from the high pressure line. The load was varied by adjusting the supply of water, there being a fixed discharge down the centre of the drill shaft to lubricate the drill proper.

A better version of this drill was developed over the next few years, in which the Pelton wheel was replaced by a reaction turbine and the used water returned to the pump by means of a coaxial hose. Both for the construction of the drill body and for the braiding of the flexible hose by which it was coupled to the high- and low-pressure mains Kellow used Aluminium-Silicon alloys, being one of the first to adopt these lightweight materials that had only become a commercial reality in the mid-1890's. A few years earlier Aluminium had been a precious metal comparable to Platinum! This drill was an engineering masterpiece. Complete it weighed little more than a hundredweight, yet the turbine developed 55 hp and it was capable of drilling a 2.5 in hole in slate or rock at the rate of seven feet in ninety seconds. This performance has never been equalled by any other portable drill—the best drills available today can drive a 1.75 in hole at forty inches a minute.

That the Kellow drill never made much impact was largely due to Kellow's personality. He was a perfectionist, not a businessman, and thus quite unwilling to compromise his principles in order to make them acceptable. His drill was a huge success at Croesor because he took the trouble to persuade the rockmen that by allowing the driller to come into their chambers, where hitherto the right of the rockmen to win rock had been inviolable, they would earn more money because they had merely to mark out where they wanted holes drilled and drill two small holes at each point to attach the drill to the rockface. This done, the drill would achieve in five minutes what it took them a day or more to do by hand with a "jumper". In short, he managed to modify the traditional bargain system. Yet when he demonstrated the drill at other quarries he failed completely to put this across and thus presented his equipment as something of an expensive toy. It never occurred to him to invite the management of other quarries to visit Croesor where they could see the drill being used properly!

The Kellow drill was made in the Keldril Works which was simply the old blacksmiths' shop at Croesor given a coat of whitewash for the benefit of a local photographer. A superb series of photographs exists showing this works in its heyday and only when one has actually surveyed the remains does it become clear just how small it was and how cleverly it was photographed.

Few Kellow drills were actually made (though fortunately three have survived in various states of completeness) and one night the Keldril Works burnt down, all the machinery being declared a write-off by the assessor who visited the scene a few days later. Within a week of Kellow banking the cheque he had dug it all out and set up a successor to the Keldril Works in another part of the buildings. It is still remembered by older inhabitants as the "Insurance Job" ....

Another classic story concerns the Kellow high-speed slate planer. I found the patent drawings for this long before I heard the story from a witness of the event, and on studying them decided that there was a fundamental weakness in the design. Kellow apparently expected to reverse a table carrying a large block of slate almost instantaneously through the medium of a single cast iron pinion, and I observed that had he ever made something that corresponded to the drawings the table would have shot off the end the first time he used it. Kellow built a planer which he installed at Parc Quarry, but instead of testing it in private he invited his friends and other quarry owners to witness the first trial. The table was duly loaded with a block of slate weighing something over two tons, the cutting tools adjusted and the planer set in motion. It made one cutting stroke, reversed at high speed as intended and then there was a bang as the teeth stripped off the pinion when it tried to arrest the rearward motion of the table. The latter carried on, demolishing the rear wall of the mill and coming to rest outside!

The Parc and Croesor concern finally closed in 1930, never having been a truly successful venture. It was probably Kellow's persistence that kept it going as long as it did. The engineering world hadn't heard the last of Moses Kellow, though. In his declining years he turned once again to the design of electrical machinery and devised a dc generator that worked like an alternator, in other words it drew the current straight off the stator windings and did not therefore require a commutator which was always the Achilles Heel of large dynamos. Nobody believed this was possible, but Kellow was able to prove to Metropolitan Vickers that he was right. Sadly he was then taken ill and nothing more was done to put his ideas to the test.

Two years before his death, the Editor of the Quarry Managers' Journal found himself in need of something to fill a few pages and asked Kellow for some reminiscences. True to form, Kellow obliged with an informal autobiography that ran to twenty-four instalments. Not always accurate and tantalisingly vague in places this is nevertheless a priceless account of Kellow's activities and fills in many gaps in the history of slate mining in the Croesor district.

Keldril Works - "Part of the Machine Shop"

Moses Kellow (Photograph from the Quarry Managers' Journal)

THE

Park and Croesor Slate Quarries

COMPANY, LIMITED.

MOSES KELLOW, Manager and Secretary.

Quarries—Park & Croesor Slate Quarries.
Postal Address—Penrhyndeudraeth, North Wales.
Telegraphic Address—"Kellow, Portmadoc."
Shipping Port—Portmadoc, North Wales.

Manufacturers of

Best Blue Portmadoc "OLD VEIN" Slates.

Ordinary Slate Ridging in 3 Pieces.

PERSPECTIVE VIEW.

SECTIONAL VIEW.

Ordinary Slate Ridging in 2 Pieces.

PERSPECTIVE VIEW.

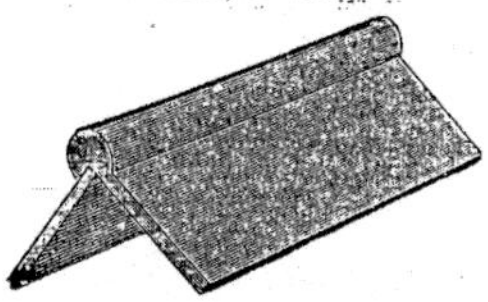

SECTIONAL VIEW.

Kellow's Patent Interlocking Slate Ridging.

WITH CREST FOR RIDGES.

WITH CREST FOR RIDGES.

THE MOST
PERMANENTLY
SATISFACTORY
RIDGING
IN ANY
MATERIAL UPON
THE MARKET.

WITH CREST FOR RIDGES.

FOR HIPS.

The Manufacture of all kinds of Slate Ridging is a SPECIALITY at these Quarries, and as the result of extended experience, the Quality and Finish obtained are acknowledged to be unequalled. Orders and enquiries for any class of Slate Ridging Work are, therefore, particularly solicited.

Slabs and all other classes of Slate Goods made to order.

THE TRADE ONLY SUPPLIED.

The Kellow Rock Drill Syndicate, Limited

KELDRIL WORKS

CROESOR : PENRHYNDEUDRAETH

THE KELLOW DRILL

TYPE K.

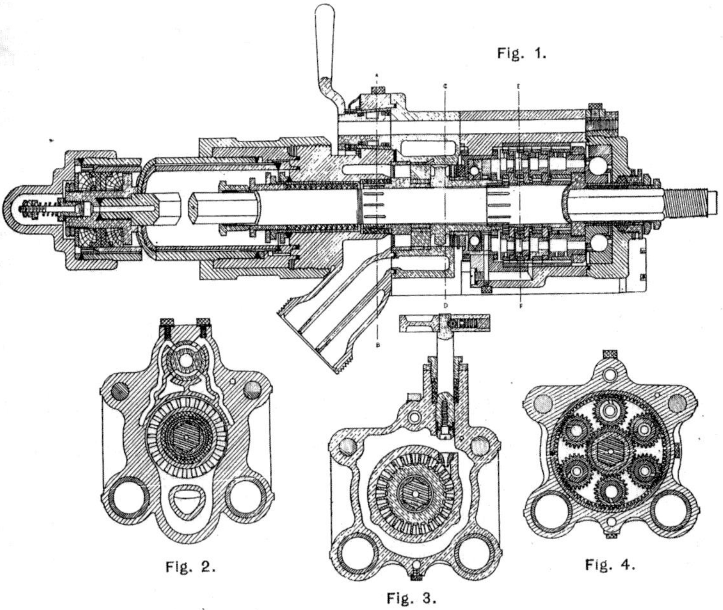

FIG. 1. Longitudinal Section.

FIG. 2. Transverse Section on line A-B of Fig. 1.

FIG. 3. Transverse Section on line C-D of Fig. 1.

FIG. 4. Transverse Section on line E-F of Fig. 1.